Self-Propelled Janus Particles

David J. Fisher

Published by **Materials Research Forum LLC**
Millersville, PA 17551, USA

Published as part of the book series
Materials Research Foundations
Volume 93 (2021)
ISSN 2471-8890 (Print)
ISSN 2471-8904 (Online)

Print ISBN 978-1-64490-118-2
ePDF ISBN 978-1-64490-119-9

Distributed worldwide by

Materials Research Forum LLC
105 Springdale Lane
Millersville, PA 17551
USA
http://www.mrforum.com

Printed in the United States of America
10 9 8 7 6 5 4 3 2 1

Table of Contents

Let me quote still another new animal: the Janus grains
Pierre-Gilles De Gennes, Nobel Lecture, 9th December 1991

Introduction

Although De Gennes used the term, animal, his lecture did not envisage that these animals would soon exhibit autonomous locomotion. Their advance has now been rapid, in every sense.

In the case of inorganic liquid solutions, diffusion is a slow random-walk process that controls redistribution of the relatively simple molecules. In biological systems, the rather complex molecules often exploit the chemical energy of their surroundings in order to be able to undergo conformational changes. One useful effect of such changes is to produce movements which are much more rapid than plain diffusion. Such so-called molecular motors have to operate at low Reynolds numbers[1]. Here the viscous forces predominate over inertial forces and the molecules are thus rather in the same position as that of a free-fall astronaut who possesses no form of rocket propulsion. Movement has to be achieved by performing cyclic conformational manoeuvres[2], as described by the seminal paper of Purcell[3]. They work at low Reynolds numbers, where inertia does not sustain motion once the driving force ceases, and movement often requires complicated chemistry in both structural and dynamic terms[4].

Attention is limited here to particles which self-propel due to reactions that are occurring on their chemically inhomogeneous surfaces, but it is interesting to note that the production of directional motion in objects which are essentially symmetrical has connections with other interesting phenomena. For example, Ernst Mach noted[5] that a body having a single hole in its surface could propel itself by alternately taking-in and ejecting a surrounding fluid. A minute glass tube which contains some gas can, when immersed in a fluid, be caused to move by acoustically stimulating the fluid[6]. Small creatures such as scallops propel themselves by performing cyclic motions which eliminate the need to use reaction-mass; thus seeming to evade Newton's third law because, although liquid swills to-and-fro, it may as well be enclosed in a sealed box. Continuing in this vein, it has even been theorized that a spacecraft which performed analogous shape-changes could exploit the very curvature of space-time for propulsion: the so-called Wisdom-Effect[7].

Back on Earth, some very simple devices in the form of chemically propelled Janus swimmers were proposed, and investigated theoretically using a coarse-grained model. That is, asymmetrical colloidal particles catalyzed the formation of products which were characterized by a poor interaction with part of the particle body. The results showed that these simple systems exhibited a propulsive motion that was comparable to those of more complicated micro- and nanoscale engines. By analogy with the behavior of macroscopic motors, the particle shape was very important and affected not only random diffusion, as in the Stokes-Einstein law, but also the propulsive efficiency. So an exciting new trend has been to imitate the molecular motors of biological systems by creating micro- and nano-scale particles which can exploit chemical energy so as to produce directional motion. These synthetic motors are subject to thermal fluctuations, and still have to operate under those conditions where viscous forces predominate. They open up many new possibilities, given that they can be caused to effect molecular transport in a similar manner to that occurring in natural biological systems. Because of their small size and controllable mobility, these micro-scale and nano-scale motors have potential applications in fields ranging from medicine (micro-surgery, *etc.*) to pollution-control (heavy-metal removal from water, *etc.*). Some of these synthetic motors have to undergo the same conformational changes as those of living creatures in order to move, but the main trend is to equip the particle with some sort of conventional third-law propulsion mechanism. Because of the very small size of the molecular motors, they can be affected by Brownian motion. This then affects the intended directionality of the motion. In the case of catalysis-propelled particles, chemical asymmetry is generally considered to be the key factor, but some studies have noted that particles having identical chemical properties can be propelled in opposite directions. It has been shown[8] that, in addition to chemical properties, the detailed shape of such a particle can play an important role.

There are two strategies for effecting motion: self-propulsion and propulsion under the influence of an external field. In the former case, the particles harvest energy from their surroundings via processes such as self-electrophoresis, self-diffusiophoresis and bubble generation. In the other case, the energy for particle motion is supplied by an electric field, acoustic field, magnetic field or light. Magnetic fields are the most promising because a low-frequency low-strength field is not only easy to create and modulate, but is also considered to be harmless to biological media. The power-transfer to the particle generally arises from an induced magnetic torque or field gradient. Directional control of the particle can be achieved by rotating or oscillating the magnetic field. Such steering methods can be applied even when the propulsion is provided by self-electrophoresis or bubble generation. The individual catalytically propelled Janus particles can self-assemble into aggregates having various translational and rotational velocities, with the trajectory of a doublet being governed by the fixed relative orientation of the particles[9]. The assembling can be controlled by using external magnetic fields to reorient the particles so as to single out those having configurations which have linear trajectories, or can be otherwise steered by an external field.

Self-propulsion

diffusiophoresis

One of several phoretic mechanisms is that of self-diffusiophoresis, in which a self-generated concentration gradient produces motion. Other possible mechanisms include self-electrophoresis and self-thermophoresis. Phoretic self-propulsion involves force- and torque-free motion at small scales. Analysis of the flow-field around a particle which is moving via self-diffusiophoresis typically neglects advection of the solute-field by the flow and assumes that the chemical interaction layer is thin as compared to the particle size. The domain of validity of the latter assumption has been determined[10] for finite values of the Péclet number. It was shown that, although advection always leads to a decrease in particle velocity and flow-stress at high values of the Péclet number, an increase could occur at intermediate values of that number. This potential increase in velocity depends greatly upon the type of chemical interaction occurring between solute and surface. Concentrating on diffusiophoresis, the means by which phoretic particles can achieve chemotaxis at both the individual and non-interacting group level have been analysed[11]. An entirely analytical law was derived for the instantaneous propulsion and orientation of a phoretic particle possessing general axisymmetrical surface properties in the limit of zero Péclet number and small Damköhler number. This was then applied to a Janus sphere, and a generalised Taylor dispersion theory was used to characterise the long-time behaviour of a population of non-interacting phoretic particles.

This is where the Janus particle comes into its own. Such particles are named after the two-faced Roman god-of-change because they also have two faces which, as it were, harbour different ambitions. They recall the familiar alkylbenzene sulfonate molecules found in detergents, where one part of the molecule is hydrophilic while the other part is lipophilic.

Particle motors can, for example, be constructed from linked spheres; one of which is catalytic while the other is non-catalytic. The asymmetrical catalytic activity then creates concentration gradients around the motor, leading to a non-equilibrium situation and consequent motion. This again recalls an even older analogy; the camphor-boat. There the dissolution of camphor in water causes a local variation in the surface energy of the water and this propels any boat-like object (viewed here as being the other half of a Janus pair) to which the camphor is attached. Study of the phenomenon dates back to the 17th century and its history was already old (figure 1) in the 19th century[12].

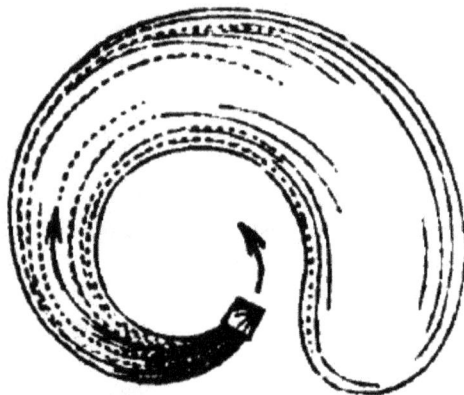

Figure 1. Motion of a camphor boat on water; woodcut from Tomlinson (1863)

As Tomlinson put it, rather poetically,

> *Every one knows that when camphor in small fragments is thrown upon the surface of clean water it moves about and rotates with considerable rapidity. This fact has long been one of the waifs of science. It has often been picked up and examined by good observers, and as often thrown aside; and as there is no scientific lord of the manor entitled to waif, it still remains unclaimed, and wanders about the regions of knowledge as a scientific outlaw.*

Colloidal particles can be set in motion by various physicochemical mechanisms. In the case of a spherical particle acting as a catalyst but exhibiting asymmetrical surface reactivity, a solute concentration-gradient can appear in the surroundings and propel the particle via self-diffusiophoresis. The solute concentration usually develops via diffusion in still environments but, in practical use, self-propelled colloidal particles might well have to function in flowing fluids. A simple shear flow, imposed on a Janus particle, can distort the self-generated solute concentration gradient and the degree of distortion is quantified by the Péclet number which is associated with the shear flow. Detailed analysis[13] of the concentration gradient surrounding a Janus particle in shear flow at a small Péclet number showed that, when the symmetry axis of the particle was aligned with the flow, the Janus particle experienced a cross-streamline drift which was of the order of the Péclet number. There was also a reduction in the translational velocity along the flow direction which was of the order of a 3/2-power of the Péclet number. The in-plane trajectory of the Janus particle in shear was such that it performed elliptical orbits around its initial position; the orbits decreasing in size with increasing value of the Péclet number. A similar theoretical investigation[14] was made of the self-diffusiophoresis of 2-faced 2-dimensional Janus particles which were being propelled by concentration gradients produced by the diffusion of a solute into the surroundings under conditions involving zero Reynolds and Péclet numbers. These model particles had piecewise-constant surface mobilities and activities over the 2 faces. The concentration gradients produced tangential boundary slip, leading to translation and rotation of the particle. An analytical solution could be found for the velocity of an isolated circular particle in free space. An explicit solution could also be found for the non-linear dynamic system of a particle confined by being situated near to a straight no-slip wall. These results showed that, if the particles did not hit the wall within a finite time, they were eventually repelled by it. A continuum model was also used[15] to simulate the confinement effect upon the diffusiophoresis of Janus particles travelling on a substrate. In experiments, a quasi 1-dimensional motion was observed in which the diffusiophoresis force predominated and the particles moved at a constant velocity along a straight line for a short interval. This was used as a reference framework for numerical study of the distributions of flow and concentration fields. Confinement had a marked effect upon the magnitudes of the forces.

The overdamped Brownian motion of a self-propelled particle, driven by a projected internal force, has been studied[16] by analytically solving the Langevin equation. The particle which was considered was limited to movement along a linear channel. The direction of its internal force was orientationally diffuse on a unit circle in a plane perpendicular to the substrate. An additional time-dependent torque acted upon the internal-force orientation. This model was relevant to particles such as catalytically-driven Janus particles or to bacteria moving on a substrate. Analytical results were found for the first 4 time-dependent displacement moments and were analysed for various special situations. With a vanishing torque, there was an appreciable dynamic non-Gaussian behaviour at finite times, as reflected by a non-vanishing normalized kurtosis of the particle displacement. This approached zero at long times, with a $1/t$ dependence.

The particles can in fact navigate along a boundary due to their hydrodynamic interactions. In a model[17] for such a process, a spherical Janus colloid was considered which was coated with a symmetrical catalyst cap that converted fuel into a solute which was then repelled from the colloid. This repulsion was assumed to occur over a distance which was much smaller than the particle radius. Within the thin interaction layer, a concentration difference developed along the surface and generated a pressure gradient which balanced the interaction force of the solute with the surface. The pressure gradient then drove a slip flow towards higher concentrations and propelled the particle away from the accumulating product. Motion near to an infinite no-slip planar wall which did not adsorb solute was analysed by solving the solute-conservation and Stokes equations. The results showed that, when the colloid was oriented with its cap axisymmetrically facing the wall, it was repelled by the accumulation of solute in the gap between the particle and the wall. When the cap was away from the wall, the particle moved towards the wall due to repulsion resulting from solute accumulation on the cap side. Very large caps accumulated solute in the gap and the particle then stopped. When a particle having a small cap-size approached obliquely, with the cap opposite to the wall, the particle was driven towards the wall by accumulation on the cap side but rotated as it approached and eventually skittered away as the cap re-oriented itself and faced the wall. When a particle having a large cap-size approached obliquely, again facing away from the wall, the product accumulation in the gap suppressed rotation and produced a normal force which caused the particle to become stationary or to parallel the surface at a fixed distance and orientation; the choice here depending upon gravity. Most theoretical analyses of the self-diffusiophoretic motion of colloidal particles have been based upon a hydrodynamic thin boundary-layer approximation and a solvent body-force due to a local self-generated solute gradient. Such a gradient tends to be introduced via the boundary conditions without considering the thermodynamic cost of maintaining the gradient. A recent approach[18] instead used a local detailed balance condition which provided a direct link between dynamics and entropy-production. In the case of self-propulsion in a de-mixing binary solvent, propulsion was attributed to forces - at the poles - that acted perpendicularly to the particle surface. More familiar results were found in the case of catalytic particles driven by the liberation of chemical free energy. In both cases, propulsion was attributed to asymmetrical dissipation rather than to a simple asymmetrical concentration of molecular solutes.

The behaviour of micron-scale autophoretic gold-platinum Janus rods, having various Au/Pt length ratios and swimming close to a wall in an imposed background flow, was investigated[19]. Their ability to move upstream, a process known as rheotaxis, depended strongly upon the Au/Pt ratio; an easily tunable variable. The numerical simulation of swimming rods actuated by surface slip revealed a similar rheotactic tunability by variation of the location of the surface slip versus surface drag. The slip location governed whether the swimmers were pushers (rear-actuated), pullers (front-actuated) or intermediate. Simulation and modelling showed that pullers exhibited a robust rheotaxis due to their larger tilt angle with respect to the wall. This made them sensitive to flow gradients. A rheotactic response intimated the nature of the difficult-to-measure flow fields of an active particle, established its dependence upon swimmer type and showed how Janus rods could be tuned so as to optimize flow responsiveness.

One example of a Janus micromotor is that of a silica particle having a platinum coating on one half. Motion is then possible due to the reduction of hydrogen peroxide on the platinum surface. It was found[20] that such particles exhibited a 3-stage behavior with regard to the dimensionless mean square displacement, while their displacement-probability distribution was double-peaked. This revealed the non-Gaussian nature of Janus-particle self-propulsion. A numerical model was used[21] to analyze experimental data on platinum-SiO_2 Janus microspheres having various shapes but a fixed volume. This showed that cylindrical and ellipsoidal Janus particles exhibit higher velocities and greater fuel consumption than do spherical particles. For a cylindrical particle, the optimum condition was that the ratio of the diameter to the length was about 0.28; this maximized both the velocity and the fuel consumption. Careful study[22] of the movement of platinum-SiO_2 particles in 2.5, 5 or 10% aqueous solutions of H_2O_2 showed that the effective diffusion coefficient of the particles was related to the time interval. For short intervals, the diffusion coefficient increased linearly with the length of the interval. Over a sufficiently long interval, the coefficient tended towards a steady value. The latter, for a given particle size, was between 10 and 100 times higher than that for pure water. This behaviour of the Janus particle was very different to that seen in Brownian motion, where the coefficient remains unchanged. There was found to exist a characteristic moment at which the coefficient ceased to increase linearly with time and instead entered a plateau stage. This characteristic time increased with increasing particle size. Its dependence upon the peroxide concentration was less clear.

Platinum is not the only choice of metal, and peroxide not the only fuel. Other spherical Janus SiO_2 particles use an iridium hemispheric layer for the catalytic decomposition of hydrazine, and here the fuel (literally 'rocket fuel') concentration can be as low as 0.0000001%; a 10000-fold decrease when compared with rival catalytic particles[23]. The iridium-SiO_2 particles also self-propel at about 20 body-lengths/s in an 0.001% hydrazine solution (figure 2), due to osmotic effects.

Catalysis-propelled mesoporous silica nanoparticles having sizes of less than 100nm are destined for use as Janus particles for the transport of 'cargoes'. Due to the mesoporous structure and small size, they can carry large quantities of guest molecules. When asymmetry of the particle was imparted by the electron-beam deposition of a 2nm platinum layer, the chemically powered Janus particles exhibited diffusion at concentrations of less than $3wt\%H_2O_2$, with the apparent diffusion coefficient being perhaps doubled with respect to Brownian motion[24].

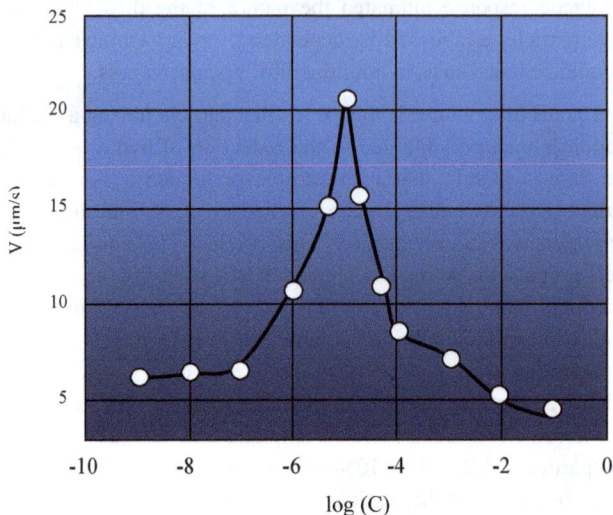

*Figure 2. Velocity of iridium-SiO₂ Janus particles
as a function of hydrazine concentration*

Janus particles of platinum-SiO₂ type, with spiropyran deposited on the SiO₂ hemisphere, exhibited[25] self-propulsion when hydrogen peroxide was present in a 1:1(vv) N,N-dimethylformamide/H₂O mixture. These self-propelled particles could dynamically assemble into groups due to electrostatic attraction and π-π stacking, induced by ultra-violet (365nm) irradiation. They also quickly disassembled when the wavelength was changed to 520nm (green) for the first time. The assembled particles could move together, propelled by hydrogen peroxide, in various patterns under ultra-violet irradiation.

Later work[26] has shown that the velocity of colloids, made from silica microspheres half-coated with platinum in solutions of hydrogen peroxide, decreases inversely with the solution viscosity. The single-particle movement of such microspheres in various viscoelastic fluids has been studied[27] by dispersing Janus colloids in dilute polyvinylpyrrolidone solution and in polyacrylamide solution in semi-dilute regimes. These environments imposed relaxation-times which ranged from some 5ms in the case of polyvinylpyrrolidone solution to about 14.5s in the case of polyacrylamide solution; both being shorter than the rotary Brownian-motion time-scale. The Janus particles tended to become physically confined by polymeric entanglements, but could still escape such cages within a time which was much shorter than the relaxation time of the polymer solution.

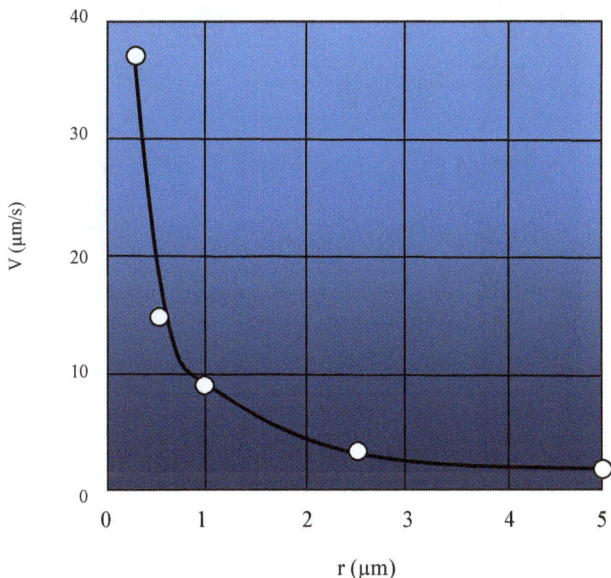

Figure 3. Velocity of platinum-polystyrene
spheres as a function of the radius

The velocity of platinum-polystyrene Janus particles driven by chemical reaction was studied[28] experimentally as a function of size. It was found that the velocity was an inverse function of diameter (figure 3). A diffusion-reaction model which assumed a 2-step process of asymmetrical catalytic activity on the surface predicted such behaviour for large particles but a plateau for smaller particles.

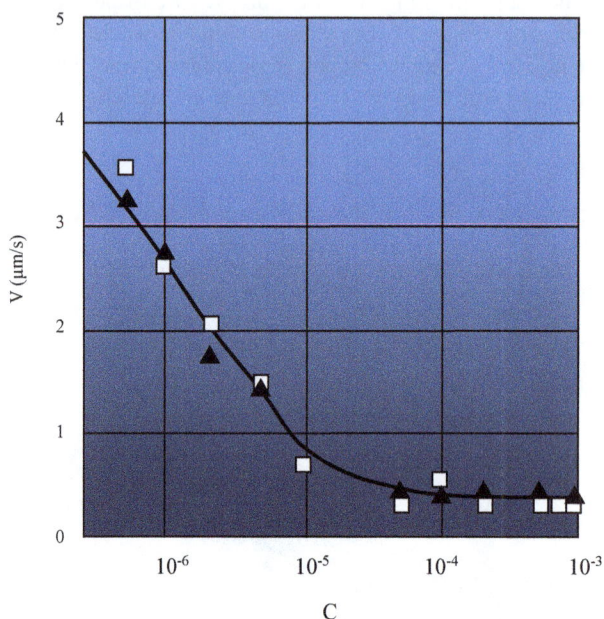

Figure 4. Velocity of 2μm-diameter platinum-polystyrene Janus particles in 10%(w/v) aqueous H_2O_2 solution as a function of $AgNO_3$ (squares) or KNO_3 (triangles) content

Study of the rotational motion showed that Brownian rotation still predominated, although the measured rotational diffusion coefficient was somewhat anomalous. It is generally accepted that neutral self-diffusiophoresis explains the propulsion of these particles, although marked ionic effects have been observed; including a reduction in speed due to the presence of NaCl. Such ionic effects were studied[29] in platinum-coated polystyrene colloids and it was found that the propulsion direction could even be reversed by adding an ionic surfactant.

Although the addition of neutral-pH salts reduced the particle speed, the addition of strongly basic NaOH had little effect. It was thus proposed that propulsion was not caused by neutral or ionic self-diffusiophoresis, but instead by the same mechanism which operates in the case of bimetallic so-called swimmers. A study[30] of the effect of salt additions upon the propulsion of platinum-polystyrene Janus colloids in hydrogen peroxide solution showed that micromolar quantities of potassium and silver nitrate salts reduced the particle velocity by similar amounts (figure 4), but had very different effects upon the overall rate of catalytic breakdown of the peroxide. This was explained theoretically by assuming a generalised sequence which involved charged intermediates and two nested reaction loops (figure 5).

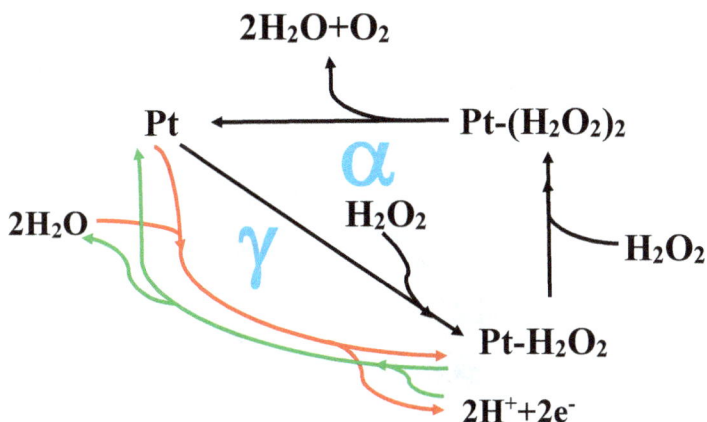

Figure 5. Catalytic reaction scheme involving coupled loops of competing reactions. Loop α: main non-equilibrium cycle comprising only uncharged species Loop γ: linked to the production of charged intermediates

Table 1. Ballistic motion of platinum-polystyrene ellipsoids in 8%H_2O_2 solution

Aspect Ratio	Platinum Area (μm^2)	Velocity ($\mu m/s$)
2.4	2.0	1.80
4.8	2.1	2.16
6.7	2.2	10.0

Table 2. Spinning motion of platinum-polystyrene ellipsoids in 8%H_2O_2 solution

Aspect Ratio	Platinum Area (μm^2)	Rotation Rate (rad/s)
2.4	2.0	2.43
4.8	2.1	2.32
6.7	2.2	4.93

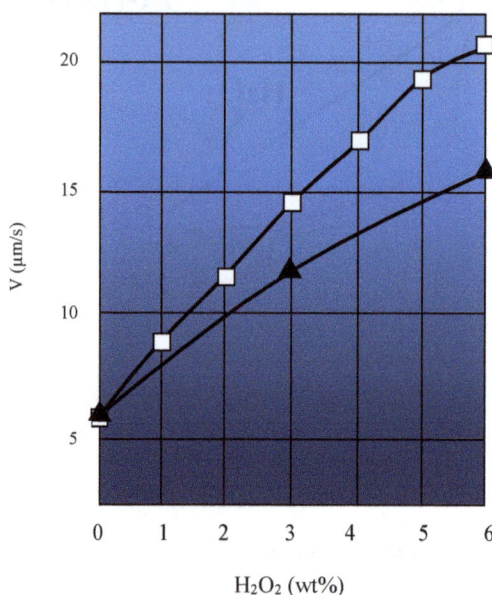

Figure 6. Velocity of aminopropyl-modified dendritic porous silica (squares) and solid silica (triangles) Janus nanoparticles (both with adsorbed NH$_2$ and 50% platinum) as a function of peroxide concentration

The propulsion of platinum-coated polystyrene prolate ellipsoids, as generated by catalytic decomposition of hydrogen peroxide, was monitored[31] via direct observation of their trajectories. These Janus ellipsoids were prepared by stretching micron-sized polystyrene spheres before coating half of the particle lengthwise with platinum along the major axis. The particles followed complicated paths in aqueous solutions containing 2 to 8%(w/v) of hydrogen peroxide. With increasing peroxide content there was a transition, from 3-dimensional passive Brownian motion to 2-dimensional active motion, which ended at 4%(w/v). The 2-dimensional trajectories of individual particles could be ballistic (table 1), where the ellipsoids moved at least 5 times more than did purely diffusive ellipsoids at the characteristic time-scale of rotational diffusion. Spinning ellipsoids moved only short distances, with a predominant rotation (table 2) about the minor axis that persisted for long periods. A third category of trajectory involved both appreciable translation and rotation. The particle velocity and active force increased with aspect ratio.

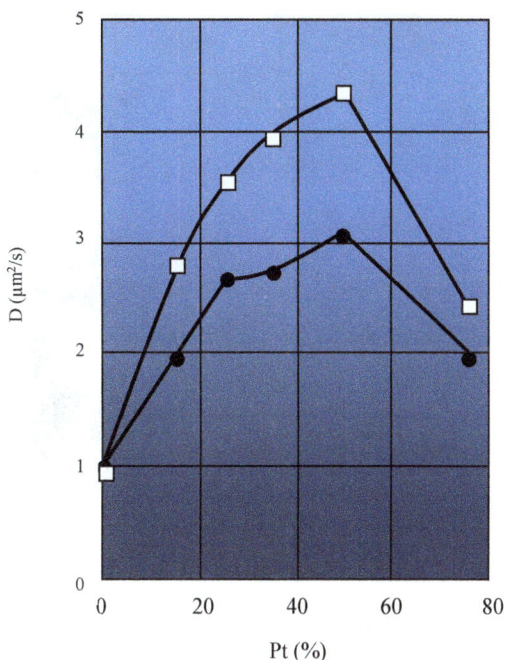

Figure 7. Diffusion coefficient of aminopropyl-modified dendritic porous silica Janus nanoparticles with adsorbed NH₂ as a function of platinum coverage in 3wt% (squares) or 1.5wt% (circles) H₂O₂ solution

Hybrid dendritic porous silica Janus nanoparticles, having coverages ranging from 0 to 100% and coatings ranging from 15 to 30nm in thickness, were prepared[32] by controlling the embedded depth of charged nanoparticles and then adsorbing oppositely charged nanoparticles. The effect of platinum coverage upon the motion of aminopropyl-modified particles in peroxide solution was studied. Particles with a diameter of 250nm and a platinum coverage of 50% exhibited the highest velocities. The velocity increased with peroxide concentration (figure 6). The effective diffusivities as a function of coverage (figure 7) and peroxide concentration (figure 8) were also determined. The propulsive effect of these particles was increased overall due to their more accessible dendritic surfaces, which enhanced the catalytic activity. It was noted that cargo delivery was possible by these particles, with a limited loss of propulsive ability.

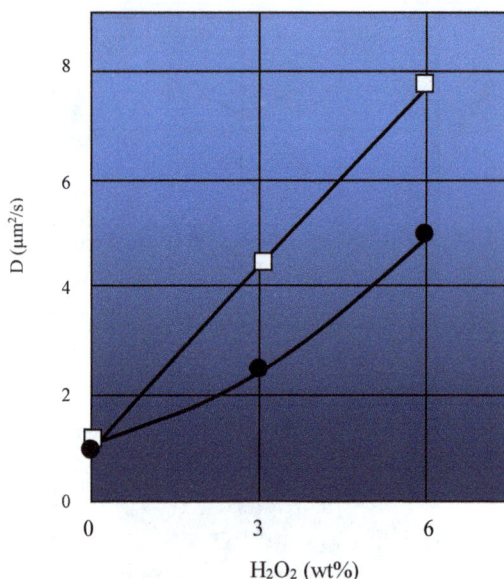

Figure 8. Diffusion coefficient of aminopropyl-modified dendritic porous silica (squares) and solid silica (circles) Janus nanoparticles with adsorbed NH_2 and 50%Pt coverage as a function of peroxide concentration

With the same design philosophy in mind, a study was made[33] of the effect of irregular surface deformations upon particle velocity. The surfaces of polymer microspheres were deformed before depositing a layer of platinum, thus resulting in the formation of nanoscopic pillars of catalyst. When exposed to hydrogen peroxide in water, these surface features more than doubled the velocity when the underlying deformation was nano in scale. Large deformations had little effect, regardless of the received catalytic surface area. Particles with deformed surfaces were also more likely, than relatively smooth ones, to exhibit a mixture of rotational and translational propulsion. It is also found[34] that the use of a simple method to control the surface roughness can produce self-propelled Janus particles, having rough catalytic platinum surfaces, which exhibit a 4-fold increase in velocity over that of particles which are coated with a smooth platinum layer.

In early work[35] on Janus particles, silica beads had different metals on opposite hemispheres; with the metal caps separated by an equatorial belt of uncoated silica. Oxidation of some of the metal components could be produced by exposure to oxygen plasma. These Janus particles can move due to ion fluxes which are driven by asymmetrical electrochemical reactions; recalling that an asymmetrical ion flux across living membranes, thus generating electric fields and producing movement, had originally been proposed for bacteria. Equations were derived[36] which treated the motion of bimetallic rod-shaped motors in hydrogen peroxide solution as being due to reaction-induced charge auto-electrophoresis. The coupled Poisson-Nernst-Planck-Stokes equations governing electrochemical reaction at the rod surface were solved numerically, and the overall analysis showed that electrokinetic motion resulted from electro-osmotic fluid slip around the nano-particle surface. The electroviscous flow was driven by electrical body-forces which resulted from coupling between a reaction-induced dipolar charge density distribution and its associated electric field. The electroviscous velocity increased quadratically with surface-reaction rate in the case of an uncharged particle, and increased linearly when the particle possessed a finite surface charge.

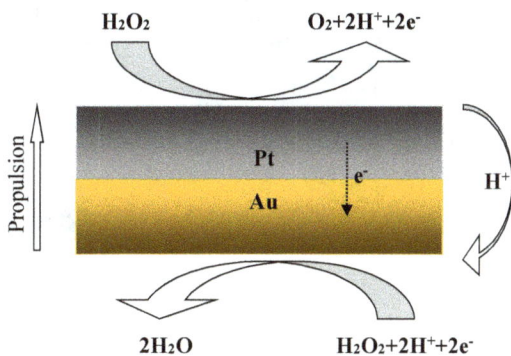

Figure 9. Janus-rod with platinum and gold plated onto polystyrene fibres

Related simulations of the particle motion showed[37] that the velocity decayed rapidly with increasing solution conductivity. These simulations again satisfied the full Poisson-Nernst-Planck-Stokes equations, assuming multiple ionic species, a cylindrical particle located in an infinite fluid plus non-linear Butler-Volmer boundary conditions which represented the electrochemical surface reactions. It was concluded that the speed of bimetallic particles was reduced in high-conductivity solutions because of reductions in the induced electric field in the layer close to the rod, in the total reaction rate and in the rod's zeta-potential. This all suggested that the auto-electrophoretic mechanism was unavoidably susceptible to speed-reduction in solutions of high ionic strength. In recent work[38], the sideways self-propulsion of straight (figure 9) and bent Janus micro-rods has been studied, with the rods being prepared by sputter-coating platinum and gold onto aligned electrospun polystyrene micro-fibers. Self-propulsion was again induced by reaction with hydrogen peroxide at the Janus-particle interface (figures 10 and 11). The self-propulsion trajectory changed from straight to circular when the particle shape was changed from straight to L-shaped. Mathematical modelling showed that the trajectory of an irregularly-shaped micro-rod depended only upon the particle shape. In earlier work[39] on the same materials, the propulsion was attributed to a potential difference.

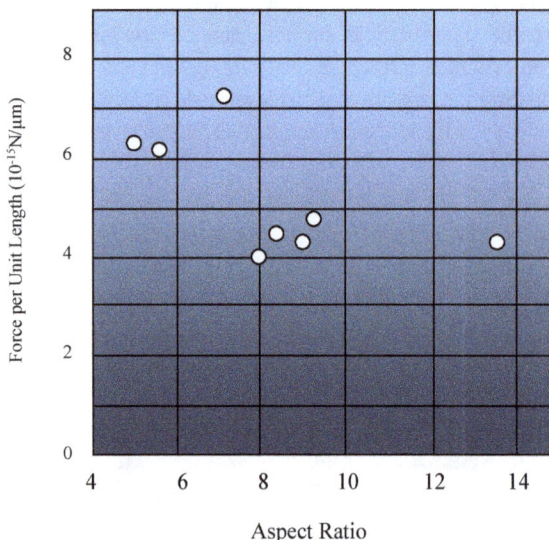

Figure 10. Experimentally estimated self-propulsion force per unit-length of straight platinum-gold Janus-rods of various aspect-ratio in 2wt% hydrogen peroxide solution

Returning to the main topic, diffusiophoresis, a dynamic model for spherical self-driven Janus particles has been combined[40] with bounce-back boundary conditions and Brownian motion, as modified by the Einstein relationship. Simulations of the self-propelled motion of 2μm spherical Janus particles were compared with experiment and the results were found to be consistent. By comparing the diffusion coefficient and diffusiophoresis force with experimental data under various conditions, the diffusiophoresis force for particles which can harvest differing concentrations of fuel was deduced. The long-range disturbance, in concentration, which is created by a reactive particle can result in a strong interaction among particles, leading to cluster formation and to localized dense and dilute regions. One simulation method[41] has incorporated the concept of so-called chemical screening, where long-range interactions are exponentially screened so that a diffusiophoretic suspension does not become unstable. Such simulations showed that uniformly reactive particles, which do not self-propel, form loosely-packed clusters. It was also found that there existed a stability threshold where, when the concentration of chemical fuel available for harvesting was sufficiently low, Brownian motion was able to overcome the diffusiophoretic attraction. The stability threshold for the clustering of Janus particles was surprisingly unaffected by their self-propulsion.

A theoretical study[42] of the self-propulsion of laser-heated Janus particles in a near-critical water-lutidine mixture related their velocity and squirmer parameter to the wetting properties of the 2 hemispheres. The hot particles moved, as usual, due to their self-generated compositional gradient, but there were 2 main mechanisms here. Dispersion forces exerted on the water and lutidine resulted in a positive velocity while, at a charged surface, the counter-ions migrated towards higher water contents and therefore drove the hydrophilic particles backwards. The first effect explained the observed forward motion of uncharged particles, and the other effect explained the backward motion of particles having charged hydrophilic caps. The velocity and the so-called squirmer parameter exhibited a non-monotonic dependence upon the wetting properties and surface temperature, thus implying that hydrodynamic interactions which were related to the squirmer parameter depended strongly upon the driving force. The behavior of squirmer systems is known to be very sensitive to the magnitude of the squirmer parameter and could be changed by modifying the heating conditions. These results suggested that an overlap of critical droplets of neighbouring Janus particles could give rise to complex interactions, and affect their aggregation behavior.

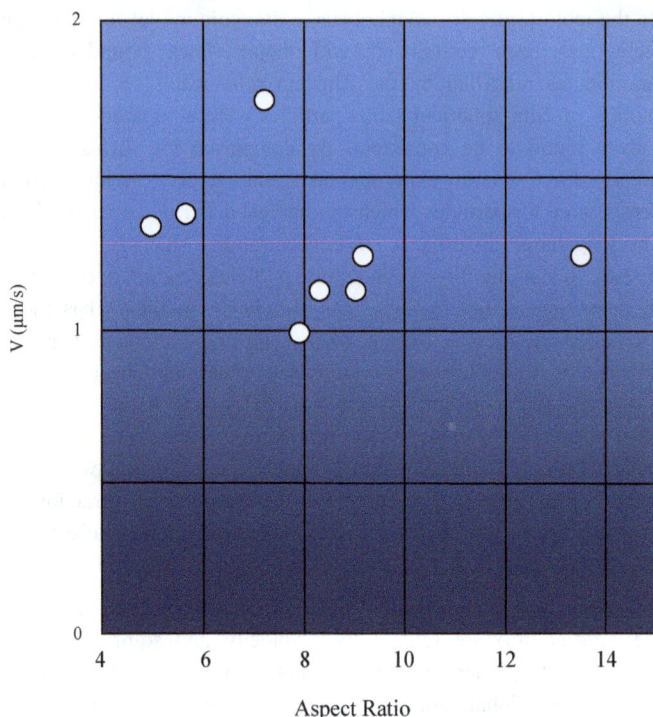

Figure 11. Experimental average velocity of straight platinum-gold Janus-rods of various aspect-ratio in 2wt% hydrogen peroxide solution

Due to the combined effects of anisotropic interactions and activity, mobile Janus particles are capable of self-assembly to form a wide range of structures; many more than those available to their more static counterparts. This can lead to the development of novel active materials which are capable of performing tasks without requiring any central control. The design of such materials requires an understanding of the fundamental mechanisms via which these swimmers self-assemble. A study was made[43] of quasi 2-dimensional semi-dilute suspensions of two classes of amphiphilic spherical swimmers whose directions of motion could be tuned. That is, they could be swimmers moving in the direction of a hydrophobic patch, or swimmers moving toward the hydrophilic side. In both cases, the hydrophobic strength of the swimmers was systematically tunable. It was observed that the anisotropic interactions, which were characterized by the angular attractive potential and its interaction range, competed with the active stress which, pointing toward or away from the attractive patch, gave rise to a rich aggregation phenomenology.

The collective effects of interacting micro-particles are largely determined by the so-called squirmer parameter, which is defined by its role in the expression for the slip velocity. Thus a so-called puller particle is propelled by the activity of its front hemisphere and has a positive squirmer parameter, while a pusher particle is propelled by the activity of its back hemisphere and has a negative squirmer parameter. At the onset of self-propulsion, the active area is just a small spot at the point of the cap and the squirmer parameter is equal to 5. For a moving particle having the active spot at the back, the squirmer parameter is equal to -5. With increasing driving force, the squirmer parameter rapidly decreases and depends markedly upon the reduced temperature. When the 2 hemispheres have opposite affinities, this can result in puller or pusher particles of variable strength. A singularity occurs where the particle velocity changes sign, meaning that a minute change in the driving force could markedly change the squirmer parameter and therefore the collective behavior. Lattice Boltzmann simulations were used[44] to study the hydrodynamics of a spherical particle near to a no-slip wall. A computational model was derived for a Janus particle by allowing for different independent mobilities on the two hemispheres, and this was compared with the behaviour of a squirmer model. The topology of the far-field hydrodynamics of the Janus particle was similar to that of a standard squirmer model, but the near-field hydrodynamics differed. With regard to the interaction between a particle and a flat wall, a numerical comparison was made of the behaviours of a Janus particle and a squirmer near to a no-slip surface. There was generally good agreement between the models, but there were some key differences for low values of the squirmer parameter. In particular, the tendency of the particle to become trapped at a surface increased for Janus particles as compared with standard squirmers. When the particle was trapped at the surface, the velocity parallel to the surface exceeded the velocity in the bulk, and was a linear function of the absolute value of the squirming parameter. A squirmer model self-propels by generating surface squirming velocities, but the mode of motion of a dumb-bell squirmer remains unclear. An investigation[45] of the swimming behavior of such a dumb-bell squirmer began with analysis of the far-field stability by means of linear stability methods. It was found that stable forward motion could not be achieved by a dumb-bell squirmer in the far field without the intervention of an external torque. The motion of a dumb-bell squirmer which was connected by a short rigid rod was treated by using a boundary-element method, as was the stability of motion of a dumb-bell connected by a spring. It was demonstrated that stable side-by-side movement could be achieved by pullers. When the rear squirmer

Materials Research Forum LLC
https://doi.org/10.21741/9781644901199

was a strong pusher, the front and rear motion was stable and the velocity greatly increased.

As mentioned earlier, there exists some disagreement concerning the exact manner in which chemical energy is transformed into motion by these particles. Some experimental observations are explained by self-electrophoresis while others are attributed to self-diffusiophoresis. There is increasing evidence that self-electrophoresis makes the predominant contribution to the motion of half-coated particles, even when most of the reaction products are electroneutral. It has been posited[46] that it might be necessary to look beyond surface reactions and instead consider the entire medium to be an active fluid which can create and annihilate charged species. The potentially predominant effect of confinement was illustrated by the case of an electro-osmotic pump which drove fluid along a substrate. Its detailed analysis permitted identification of an electro-osmotic driving mechanism which was powered by micromolar salt concentrations. Numerical methods, based upon the lattice Boltzmann method, were also anticipated to permit study of the collective behavior of systems containing mobile particles. The lattice Boltzmann method is a mesoscopic technique which is based upon the microscopic particle characteristics of the fluid. This gives a greater feeling for the interaction between fluid and solid and is better than familiar numerical methods for the description of interface dynamic problems such as the self-propulsion of Janus particles. On a certain time-scale, when a Janus particle undergoes directional motion, the effect of Brownian forces can be neglected and the analytical process simplified. Momentum exchange has been introduced into the half-way bounce-back step of the lattice Boltzmann method in order to model[47] multicomponent diffusion and reaction. By taking account of the surface force, the diffusiophoresis acting upon a Janus particle could be deduced. Analysis of the variation of diffusiophoresis proved that the magnitude of the diffusiophoresis is independent of fluid velocity. The results further showed that self-diffusiophoresis is governed mainly by the axial projected area and that the reaction area of the particle affects diffusiophoresis.

In order to explore the question of the exact process involved in ionic self-diffusiophoresis, a dielectric-AgCl Janus particle was created[48] which clearly moved away from the AgCl side when it was exposed to ultra-violet, or powerful visible, light. Numerical simulation and acoustic levitation experiments provided some explanation for a decrease in its speed over time, and for its directionality. Photo-active AgCl micro-particles also exhibited interesting gravitactic phenomena, which betokened 3-dimensional transport. In the natural world, phototactic micro-organisms which are exposed to natural sunlight respond by swimming upwards against gravity. Synthetic photochemically-active particles can also move contrary to the gravitational attraction.

They initially settle but, when subjected to low light intensities, they perform wall-bound movements near to the bottom surface. With increasing light intensity, the particles leave the wall and move away from the light source and against gravity. A theoretical model[49] which was based upon self-diffusiophoresis revealed that photochemical activity and the phototactic response were key mechanisms. Because the threshold intensity for lift-off depends upon the particle size, it can be used to affect only those particles having a given density among a diverse mixture of active particles.

thermophoresis

It was noted that colloidal Janus particles, having metallic and dielectric faces, moved rapidly when illuminated by defocused optical tweezers, without requiring any chemical fuel[50]. Instead of exhibiting random motion, the optically-activated particles circulated through the light beam and followed the rosette curves typical of hypotrochoids. This behaviour was attributed to a combination of self-thermophoresis and optically-induced torque.

The self-thermophoretic velocity of a light-irradiated spheroidal Janus nanoparticle, consisting of symmetrical dielectric and perfectly conducting semi-spheroids, was predicted[51] by solving the linearized Joule-heating problem due to uniform laser irradiance and by determining the temperature fields inside and outside the particle. Thermoelectric methods were used to find the surface self-induced temperature gradient and slip velocity which controlled the self-propulsion of the Janus particle. Simplified expressions were found, for the self-thermophoretic velocities of prolate and oblate Janus particles, in terms of their aspect ratio.

The flow-field has been measured[52] around a pinned micron-sized self-thermophoretic Janus particle which was confined within slit pores of various widths. The flow-field was disclosed with the help of gold nanoparticles. These were thermophoretically inactive and therefore did not respond directly to the generated temperature gradients. The flow-fields were altered when the slit pore was narrowed and revealed clear signs of thermo-osmotic flows which were generated by the temperature gradients that existed along the confining glass cover-slip surfaces. It was concluded that many observations of active and passive particle-clustering around phoretic swimmers were affected by osmotic creep flows which were introduced at the substrate boundary.

The motion of a Janus particle within the inhomogeneous temperature field generated by an optically heated gold nanoparticle has been studied[53], where the latter particle consisted of a polystyrene sphere which was coated on one hemisphere with a 50nm gold film. The Janus particle was held within the ambit of the immobilized gold nanoparticle by photon-nudging; which propelled the Janus particle towards a target. When close to the heat source, propulsion was switched off. An angle-dependent repulsion of the particle away from the heat-source was noted. Angular motion of the Janus particle further resulted in polarization of the Janus particle within the temperature field.

Fuel-free motion of spherical magnetic Janus particles, avoiding the usual catalytic chemical reactions and toxic reagents, could be produced[54] by magnetically-induced thermophoresis due to a local temperature gradient. The alternating-current magnetic field drove thermophoretic motion via induction-heating of a magnetic cap on the particle. A superposed direct-current magnetic field further served meanwhile to orient the Janus particles and guide their motion due to the particular properties of 100nm-thick Permalloy films which supposedly generated a topologically stable magnetic vortex state within the cap structure of the Janus particle. The method was also considered to be superior to optically-induced laser-heating because it did not require transparent surroundings. In the latter method, self-propulsion of a half metal-coated colloidal particle was caused by the absorption of intense light by the metal, thus creating a local temperature gradient which then propelled the particle via thermophoresis. The temperature distribution and thermal slip-flow field around a microscale Janus particle could in fact be measured[55], and the temperature drop across the particle, together with the velocity, agreed with predictions. In a further refinement, the motion of a laser-heated Janus particle under a rotating electric field was measured experimentally[56]. In the low-frequency range of 1 to 6kHz, circular motions of the Janus particle were observed which could either follow or counter the direction of the rotating field, depending upon the direction of electrorotation. At frequencies above 10kHz, pure electrorotation and electrothermal flow were alone observed. It was proposed that the tangential component of circular motion was caused by an electric field-enhanced self-thermophoresis which was proportional to the laser heating-power and the electric field strength. This suggested that thermophoresis could be modified by the induced zeta-potential of the Janus particle when tuned by the applied electric field. The intrinsic thermophoresis could thereby be increased several-fold at an applied voltage of only about 3V.

Langevin equations have been derived[57] for the self-thermophoretic dynamics of Janus particles which are partially coated with an absorbent layer heated by a radiation field. The derivation was based upon fluctuating hydrodynamics and radiative heat-transfer, and the use of stochastic equations for the bulk phases and surface processes which were consistent with microscopic reversibility. Expressions were obtained for the self-thermophoretic force and torque, under arbitrary slip boundary conditions. Over-damped Langevin equations for colloid displacement and radiative heat transfer furnished expressions for the self-thermophoretic velocity. A non-equilibrium fluctuation formula further showed how the probability density of Janus-particle displacement and radiation energy-transfer during a given time-interval are related to mechanical and thermal properties which characterize the non-equilibrium system state.

The force and torque on a Janus particle, moving in a rarefied gas under a thermal gradient, have been calculated[58] for a high Knudsen-number regime. The momenta of impinging gas molecules were based upon a Chapman-Enskog distribution or a binary Maxwell distribution between opposing parallel plates at differing temperatures. The reflection properties of the Janus-particle surface were characterized by accommodation coefficients which took constant but dissimilar values on each hemisphere. It was found that the Janus particle preferentially oriented itself so that the hemisphere having the larger accommodation coefficient pointed towards the lower temperature.

Locally-heated Janus colloids can move through an electrolyte due to self-thermo(di)electrophoresis. Analytical and finite-element numerical studies were made[59] of the self-propulsion of colloids in a monovalent electrolyte, taking account of electrostatic screening for intermediate and large Debye-lengths and of the fluid flow generated by self-thermo-electrophoresis. Excellent agreement was found between the analytical and numerical results in the limit of high salinity. At low salt concentrations, Teubner's integral formalism yielded expressions, for the velocity, which agreed semi-quantitatively with numerical results for conductive particles. This supported the rather high predicted velocities at very low ionic strengths which had been found numerically for entirely non-conductive particles.

Hybrid micro- and nano-motors having several separate propulsion modes are expected to exhibit improved motion in complex body fluids. A multiple-stimuli propelled lipase-modified dendritic silica/carbon-Pt particle having built-in motors, for hybrid propulsion by H_2O_2, light or enzyme, exhibited[60] increased motion under stimulus by H_2O_2. This produced an oxygen concentration gradient which arose from the asymmetrical catalysis of platinum nanoparticles. When irradiated with near-infrared light, an uneven photothermal effect of the carbon portion propelled the nanomotor via self-thermophoresis. Lipase was also efficiently loaded into the dendritic pores. This decomposed triglyceride on the silica part and led to self-diffusiophoretic propulsion.

Although thermophoresis is a common mechanism which can drive the autonomous motion of Janus particles in suitable environments, the interaction of particles with the substrate beneath the particle has remained unclear. The effect of a poly(N-isopropylacrylamide) functionalized substrate, having various chain-lengths, upon the active motion of a single polystyrene particle half-coated with gold was explored[61]. It was shown that modification of the substrate with polymer brushes increased the particle velocity; with the brush chain-length playing an important role. The results revealed an intrinsic dependence of the particle velocity upon the flow boundary condition and thermo-osmotic slip at the interface.

electrophoresis

The assembly of Janus particles by alternating-current electric fields at frequencies above 10kHz was used early on[62] in order to investigate the relationship between the field-induced dielectrophoretic force, the field distribution and the structure of the assembly. The particle concentrations were large enough to form a monolayer on a glass surface between two gold electrodes, and a wide variety of metallodielectric particle structures and behaviours was observed. These were quite different to those obtained via the directed assembly of ordinary dielectric or conductive particles using fields of similar frequency and intensity. The metallodielectric particles formed new types of chain structure in which the metallized halves of neighboring particles were aligned in lanes along the direction of the electric field while the dielectric halves faced in alternating directions. Later work on polystyrene Janus particles having gold and dielectric multilayer coatings on one hemisphere showed[63] that this markedly altered the dielectrophoresis behavior of homogeneous precursor particles in alternating-current electric fields. An alkanethiol coating on the gold-coated hemisphere could modify the dielectrophoresis cross-over-frequency; with a negative to positive transition occurring with increasing alternating-current frequency. All of the measured cross-over frequency results could be collapsed onto a single curve by scaling with the resistance-capacitance time of the alkanethiol layer capacitance and with the conductive medium resistance. A method was presented for producing large numbers (more than 10^6 particles/ml) of Janus particles[64]. Gold-(fluorescent)polystyrene particles were then prepared and a flip/flop rotational effect was noted in a microfluidic channel, due to dielectrophoresis[65].

The non-linear electrophoretic transport of uncharged and ideally-polarisable Janus spheres was considered[66], where inhomogeneity was produced due to variable Navier slip conditions at the particle surface. A general 3-dimensional model permitted the calculation of the electrophoretic mobility of any patchy particle under arbitrary slip boundary conditions. The particular case of a Janus sphere which consisted of 2 equal hemispheres, exhibiting a constant but differing slip boundary condition, was solved. In the case where the slip coefficients for each hemisphere were equal, an induced-charge electro-osmotic flow became evident at an increased rate when compared with an homogeneous sphere exhibiting no slip. If the slip coefficients differed, the particle self-aligned with the electric field and moved with the slip surface facing forwards. Analytical and numerical solutions were also provided[67] for the non-linear induced-charge electrophoretic movement of an electrically inhomogeneous Janus sphere which consisted of 2 hemispheres having differing dielectric permittivities and which was subjected to a uniform alternating-current electric field. The analysis remained valid even when the particle radius was of the same order as the electric double layer thickness. It was deduced that there were critical values for the conductivity of each hemisphere and for the frequency of the applied field which, if exceeded, could cause the mobility to decrease rapidly to zero. Analytical and numerical solutions were given[68], in the limit of uniform direct-current electric forcing, for the linear velocity and angular rotation of a single Janus particle suspended in an infinite medium. It was concluded that particle mobility was a function of the permittivity of each hemisphere, and of the difference between them, as well as of the electric double layer length. When both hemispheres had a finite permittivity, the maximum mobility and rotation were not given by the Helmholtz-Smoluchowski thin electric double layer limit but were functions of the permittivity and electric double layer properties.

It was shown experimentally[69] that an alternating-current electric field was an effective means for suppressing Brownian motion and for controlling the direction of self-propelled particles. The self-propulsion and dielectrophoretic response of a 2µm Janus particle was monitored by using pulsed 20µm-wide interdigital electrodes to modulate its self-propulsion. This resulted in a single Janus particle executing to-and-fro movements within the strip electrode. The change in direction depended upon its position: the catalyst side was always pointed outwards, and the orientation relative to the electrode was about 60°. An attempt was made[70] to create platinum-gold Janus particles having an overall size which was commensurate with that (~30nm) of an enzyme. The resultant nanoparticles could move via self-electrophoresis. The geometrical anisotropy of the Janus nanoparticles permitted the simultaneous observation of translational and rotational motion using dynamic light-scattering. It was concluded that, although strongly influenced by Brownian rotation, the Janus particles exhibited sufficient linear ballistic motion to produce increased diffusion.

Molecular dynamics and lattice-Boltzmann simulations have been used[71] to study the behaviour of charged Janus particles in an electric field. For a relatively small net charge and thick electrostatic diffuse layer, the mobilities of Janus particles and of uniformly charged colloids were identical; for a given net charge. In the case of higher charges and thinner diffuse layers, the Janus particles always exhibited a lower electrophoretic mobility. The Janus particles also aligned with the electric field, and any angular deviation from the field direction was related to their dipole moment. The latter was affected by the thickness of the electrostatic diffuse layer and was closely related to the electrophoretic mobility.

The motion of metallodielectric Janus particles moving between 2 parallel electrodes due to electrophoresis was studied[72]. A constant voltage was used repeatedly to charge conductive particles within a dielectric fluid and thus cause rapid oscillatory movement between the electrodes. In addition to oscillating, micron-scale Janus particles also moved over large distances, perpendicularly to the field, at speeds of up to 600μm/s. This behaviour was attributed to rotation-induced translation of the particle, following charge-transfer at the electrode surface. This mechanism was supported by experiments using fluorescent particles, which signalled their rotational motion and by simulations which accounted for the relevant electrostatic and hydrodynamic effects. Interaction among a number of particles could lead to repulsion, attraction and/or cooperative motion, depending upon the position and phase of the oscillating particles.

A detailed theoretical study has shown[73] that electrokinetic effects can play a role even in the motion of metal-insulator spherical Janus particles. It was noted the reaction rates depend upon the thickness of the metal coating, and that this varies from the pole to the equator of the coated hemisphere. Motion is then due to a combination of neutral and ionic-diffusiophoretic, as well as electrophoretic, effects and their interplay can be modified by altering the ionic properties of the surrounding fluid.

The direct-current dielectrophoretic manipulation of polystyrene-based Janus particles in microchannels was investigated numerically[74]. A small potential difference was applied across the microchannel, but a highly non-uniform electric-field gradient was nevertheless generated and the particles encountered dielectrophoretic forces when flowing where the highest field-gradient existed. By adjusting the conductivity of the surrounding medium, some particles could be caused to experience a negative force while others experienced a positive force. It was thus found to be feasible to separate 5μm Janus particles from homogeneous polystyrene particles, as well to separate 3μm from 5μm Janus particles. The results also showed that Janus particles having gold coverages of greater than 50% would experience a positive dielectrophoretic force. The effect of the gold coating-thickness upon particle trajectories could be neglected here.

Induced-charge electrophoresis behavior was investigated[75] for magnetic-metal coated Janus particles in which the particle orientation with respect to the electric field was controlled by an external magnetic field. In the case of a 10nm metallic coating, the particle generally moved normal to the electric field, with no magnetic field. The velocities of induced-charge electrophoresis in various directions were measured in a magnetic field, and orientation-dependent velocities were clearly observed. In the case of a 70nm metallic coating, the particle generally moved parallel to the electric field, with no magnetic field, and exhibited an opposite tendency for orientation-dependent velocities when compared with thinly metal-coated particles. It was deduced that the

Materials Research Forum LLC
https://doi.org/10.21741/9781644901199

strength of the induced dipole of a Janus particle depended upon the thickness of the metallic coating, thus resulting in differing orientation-dependent velocities of Janus particles in electric fields.

A continuum-theory study was made of the effect of association-dissociation reactions on the self-propulsion of colloids driven by surface chemical reactions[76], showing that association-dissociation should indeed exert a large influence on particle behaviour and thus be accounted for when modelling. It was noted that such bulk reactions could permit charged particles to propel themselves electrophoretically, even if all of the species involved in the surface reactions were neutral. The reactions could also increase the predicted speed, of particles which were propelled by ionic currents, by up to an order of magnitude. When the surface reactions produced both anions and cations, bulk reactions introduced an additional reactive screening length which was analogous to the Debye length of electrostatics. This then led to an inverse relationship between particle-radius and velocity. Also considered was the effect of the Debye screening length itself, showing that the usual approximation failed in the case of nanoscale particles.

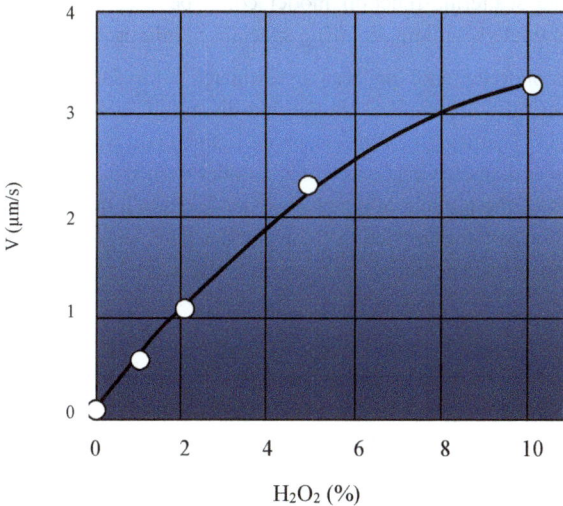

Figure 12. Velocity of platinum-silica Janus particles, having a diameter of 1 to 2μm, as a function of H₂O₂ content

Theoretical consideration[77] of the dynamics of a self-propelled Janus particle moving in an external electric field indicated that the field could affect the trajectory of the particle and direct it towards a given target. The orientational dynamics of the Janus particle behaved mathematically like a pendulum, with an angular velocity which was sensitive to both the electric field and to the surface activity of the particle.

Experimental characterisation of the swimming statistics of populations of micro-organisms or artificially propelled particles is essential to the understanding of the physics of active systems and their exploitation. A theoretical framework was constructed[78] for the extraction of information on the three-dimensional motion of micro-swimmers on the basis of intermediate scattering function data arising from differential dynamic microscopy. Theoretical expressions were derive for the intermediate scattering functions of helical and oscillatory breast-stroke swimmers, and the theoretical framework was tested by applying it to video sequences of simulated swimmers having precisely-controlled dynamics. The theory could be applied to the experimental study of helical swimmers such as active Janus colloids or to suspensions of motile micro-algae. It was shown that the fitting of differential dynamic microscopy data to a simple non-helical intermediate scattering-function model could be used to derive 3-dimensional helical motility parameters without requiring specialised microscopic equipment.

So-called dielectrophoretic choking was investigated[79] for Janus particles that were moved electrokinetically through a converging-diverging microchannel by a direct-current electric field. The negative dielectrophoretic force that would presumably block a particle which had a diameter that was much smaller than that of the inlet was in fact relaxed by rotation of the particle in a sense which minimized the force and allowed the particle to pass through the constriction. Numerical results showed how Janus particles having non-uniform surface potentials overcame the dielectrophoretic force.

The use of electric and magnetic fields to control the direction and speed of induced-charge electrophoresis-driven metallic Janus microrobots was described[80]. A direct-current magnetic field which was applied in a direction perpendicular to the electric field constrained the linear movement of particles to a 2-dimensional plane. Phoretic force spectroscopy, a phase-sensitive method which detects the motion of phoretic particles, was used to characterize the frequency-dependent phoretic mobility and drag coefficient of the phoretic force. When the electric field was scanned over a frequency range of 1kHz to 1MHz, the Janus particles exhibited an induced-charge electrophoresis-direction reversal at a crossover frequency of about 30kHz. When below this frequency, the particle moved in the direction of the dielectric side of the particle. When above this frequency, the particle moved towards the metallic side. The induced-charge electrophoresis drag coefficient was similar to that of the Stokes drag.

bubble generation

It was shown previously that, in the case of platinum-coated silica Janus microspheres, the platinum side acts as a catalysis which decomposes H_2O_2 solution and leads to self-propulsion. Some experiments[81] showed however that, for a given peroxide concentration, microspheres having a diameter of about $1\mu m$ exhibit self-diffusiophoresis propulsion while those having a diameter of about $20\mu m$ exhibit bubble self-propulsion. There are also marked differences in velocity (figures 12 and 13) and trajectory. A numerical model based upon a simplified force-balance analysis, velocity-field distributions and oxygen concentration distribution around the microspheres could explain the positions and sizes of the bubbles which were generated. It was deduced that the wall-slip coefficient was a key parameter, and that 2 slip-coefficients having an order of magnitude difference corresponded to the 2 different types of self-propulsion.

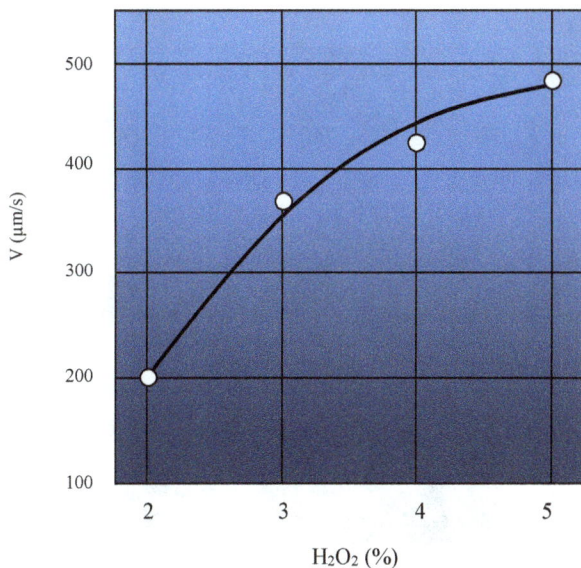

Figure 13. Velocity of platinum-silica Janus particles, having a diameter of 20 to 40μm, as a function of H_2O_2 content

Further study[82] of the motion of particles smaller than 5μm showed that it was dominated by self-diffusiophoresis, while the time-variation of the dimensionless mean-square displacement indicated that the particles exhibited simple Brownian motion at short times, self-diffusiophoresis at intermediate times and Brownian-like motion again at long times. Rotation of the Janus particles was governed by Brownian torque, and was influenced by solid walls and shear flow. When the particle size was 20 to 50μm, microbubble propulsion was of course observed. On the basis of the particle motion, 3 typical stages were identified: self-diffusiophoresis, microbubble growth and bubble collapse. Three different scaling laws (figure 14) described the microbubble growth, reflecting its control by viscous forces, capillary forces and ambient fluid pressure as the bubble size increased. A microjet was observed to form during the microbubble's asymmetrical collapse, and this could propel the particle at speeds of up to 0.1m/s. It was also revealed that an alternating-current electric field could induce dielectrophoresis and thereby influence particle motion.

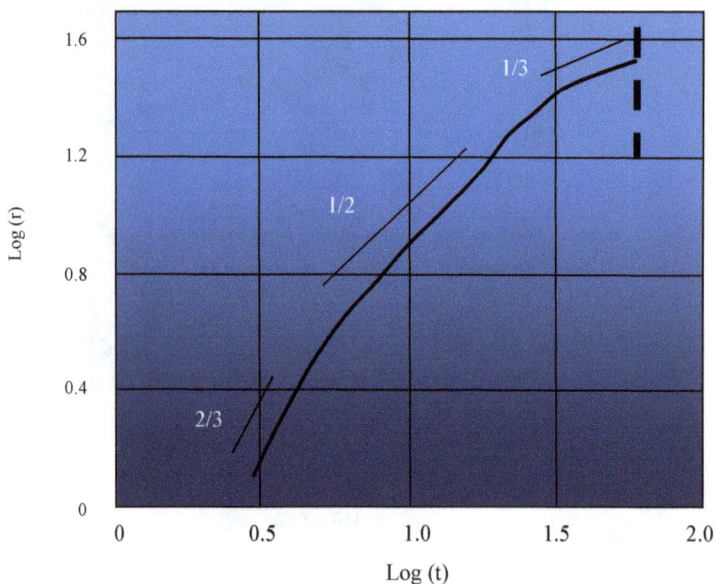

Figure 14. Various scaling laws governing microbubble growth on platinum-silica Janus particles in 2%H_2O_2 solution. The fractions indicate the power-law exponents of the time-dependence, and the vertical dotted line indicates bubble-collapse

Silica is not the only choice available. The decomposition of H_2O_2 is also triggered by a platinum catalytic layer which is asymmetrically deposited on mesoporous ZnO microparticles[83]. Those having an average size of about 1.5µm exhibit enhanced self-diffusiophoretic motion, while fast bubble-propulsion is exhibited by particles larger than 5µm. In spite of their density, they can in fact exhibit unusually high speeds of more than 350µm/s in H_2O_2 concentrations lower than 5wt%. The high speed is attributed to the efficiency of bubble nucleation and growth due to the highly active and rough surfaces. Particles having a smooth surface and low surface area remain motionless. The interrelated rapid movement and high catalytic performance of these particles suggest that they would be useful for the removal of dyes and nitro-aromatic explosives from contaminated water. In this connection, Janus particles can be usefully employed as carriers of indicator-materials. A luminescent sensor for 2,4,6-trinitrotoluene (TNT) has for example been based[84] upon so-called Janus capsules. The latter were prepared via the layer-by-layer assembly of functionalized polyelectrolyte microcapsules, followed by the sputtering of a platinum layer onto one half of the capsule. As usual, the capsules were bubble-propelled at speeds of up to 110µm/s. The detection of TNT was thereby greatly improved because the motion of the Janus capsules increased the probability of collision with a TNT molecule. This led to detection limits which could be as low as 2.4ng/ml of TNT within 60s.

Other functionalized Janus particles of various sizes have been created[85] by asymmetrically loading alginate hydrogel beads with Prussian Blue. Their synthesis was possible by using an electric-field based symmetry-breaking approach and ionically cross-linked alginate beads. This led to an oscillatory dynamic behavior of the particles which was coupled to chemical light-emission because the Prussian Blue acted as a catalyst and drove both light emission and oxygen production in the presence of luminol and hydrogen peroxide, while the differential porosity distribution of the hydrogel particle meanwhile led to the asymmetrical release of oxygen bubbles which propelled the particle.

In another example of pollution-control, magnetically steerable self-propelled particles were used to remove radioactive cesium alone from contaminated water[86]. In order to achieve this, mesoporous silica microspheres were functionalized with copper ferrocyanide, and half of the particle surface was coated with ferromagnetic nickel and catalytic platinum layers so as to create Janus particles. The latter executed random movements in H_2O_2 solution, due to catalysis-driven evolution of bubbles from the platinum surface. These self-propelled 'sponges' achieved an 8-fold higher cesium removal-rate, over 1h, as compared with that of a stationary adsorbent material. The ferromagnetism of the nickel layer permitted the particle to be magnetically steered, and

Materials Research Forum LLC

https://doi.org/10.21741/9781644901199

its speed under the influence of the field was also some 11 times higher than that without the field. Even in the presence of 1000ppm of Na^+ ions, more than 98% of the radioactive ^{137}Cs ions were removed from solution. The ability to interact with ions in that way also opens up the possibility of interacting with the much larger targets represented by bacteria; the latter were, after all, the original inspiration for the development of self-propelling particles. A so-called motion-capture-lighting strategy has been proposed[87] which combines the motion-enhanced capture of bacteria and the resultant emission of fluorescence. Janus fiber-rods have been prepared by the cryocutting of aligned electrospun fibers, and catalase was grafted onto one side of the fibres in order to produce oxygen bubbles for their propulsion. Mannose was incorporated on the other side in order to recognize specific proteins on a bacterial surface. Janus rods having an aspect-ratio of 0.5, 1, 2 or 4 were studied, and that ratio was found to have a marked effect upon the rod trajectory and speed. In particular, rods having an aspect-ratio of 2 exhibited much larger mean square displacements, and directional motion. This then led to greater bacterial capture and to larger changes in fluorescence intensity. Bacterial capture lit up the rods due to the aggregation-induced emission effect of tetraphenylethene derivatives. Under ultra-violet light, the fluorescence of Janus-rod suspensions changed from blue to bluish-green and green upon incubation with *E.coli* concentrations of 100 and 100,000CFU/ml, respectively. The limit of detection was predicted to be about 45CFU/ml within 60s.

Microtubular particles with platinum on concave surfaces, can self-propel due to the release of bubbles from one end, while convex Janus particles tend not to generate bubbles but instead move mainly due to self-diffusiophoresis. A simple chemical deposition method was used[88] to make platinum-polystyrene Janus dimers which were propelled by the periodic growth and collapse of bubbles on the platinum-coated part. Both a high catalytic activity and a rough surface were required in order to change the propulsion mode from self-diffusiophoresis to bubble generation. The present Janus dimers combined geometrical and interfacial anisotropy and exhibited motions, at the bubble growth and collapse stages, which differed by 5 to 6 orders of magnitude in time.

In view of the expense of platinum, alternative self-propelled nano-particles have been based upon the TiO_2/MnO_2 Janus combination[89]. Nanoflakes of MnO_2 are here grown *in situ* on one hemisphere of a TiO_2 sphere via the photo-reduction of $KMnO_4$ under aerobic conditions. The MnO_2 nanoflakes catalytically decompose hydrogen peroxide so as to generate oxygen bubbles which then produce movement through the solution. Such particles can be created *en masse* by using phase separation and ultra-violet induced monomer polymerization techniques[90]. By adjusting the volume ratio of 2 immiscible oils, ethoxylated trimethylolpropane and triacrylate/paraffin oil, in an initial emulsion the geometry of the resultant particles can be varied from nearly spherical, to hemispherical and to crescent-shaped. The size of the particles can also be varied by adjusting the flow of fluid. Nanoparticles of Fe_3O_4 and MnO_2 have been loaded onto Janus particles so that the Fe_3O_4 could act as a catalyst for pollutant-degradation and also control the direction of movement of the particles. The presence of MnO_2 nanoparticles on concave particles catalyzed H_2O_2 and led to bubble propulsion, further increasing pollutant-degradation.

Janus particles need not be solid. They have been prepared[91] by partially coating one side of an oil droplet with aluminium particles. Upon placing this droplet in an alkaline solution, reaction between the aluminium particles and OH^- generates hydrogen gas which is emitted from the particle-coated side of the droplet as bubbles, thus propelling the droplet in the opposite direction. Experimental studies show that, with passing time, the droplet motion exhibits 3 stages of movement: an initial development stage, a stable stage and a decline stage. This recalls the behaviour of platinum-silica particles. The speed of the droplet increases with increasing pH value and particle coverage. Surfactants produce spontaneous motion of the droplets, and their directional motion can be controlled by applying an external direct-current electric field.

Fuel-free biocompatible swimmers having dimensions smaller than 1mm could revolutionize the manipulation of microscopic systems. Sub-micron metallic Janus particles can be rapidly and autonomously propelled by acoustically-induced fluid-streaming, but are limited in utility. On the other hand, bubble-based microswimmers have an on-board resonant cavity which permits them to function when far from the source of acoustic power. Fabrication by direct writing techniques has limited their minimum dimensions and the numbers which can be produced. Size-scaling of the properties of bubble swimmers has thus not been experimentally explored and 3-dimensional autonomous motion has not been demonstrated for this type of swimmer. A method has recently been described[92] for the fabrication of bubble swimmers in large numbers, and having sizes ranging from 5μm to 500nm. These swimmers obeyed a previously proposed scaling theory and exhibited phenomena which permitted their propulsion in different manners within a given experiment: including magnetic steering.

In addition to catalysis, another method of producing propulsive bubbles is irradiation. X-rays have been used[93] to propel semi metal-coated Janus particles in aqueous solution. The particles can be both propelled and watched in real-time by means of transmission X-ray microscopy. Observations have shown that motion results from bubble growth, as increased by radiolysis of the water near to the particle surface. The speed is thus remotely controllable by varying the radiation dose. The speed of bubble-propelled Janus microcapsules with polyelectrolyte multilayers can even be varied by changing the type and concentration of a counter-ion after grafting-on salt-responsive poly[2-(methacryloyloxy) ethyl trimethylammonium chloride] brushes[94]. Reversible switching between low speeds and high speeds was made possible by changing the poly[2-(methacryloyloxy) ethyltrimethylammonium chloride brushes from hydrophobic to hydrophilic via ion-exchange with ClO_4^- and polyphosphate anions.

Janus particles which were based upon metal-organic framework materials[95] self-propelled in hydrogen peroxide due to the asymmetrical deposition of a catalytically active silver patch on the surface of the composite microspheres. The silver patches split the H_2O_2 and formed oxygen bubbles[96]. As a result, the Janus particles could attain speeds of over 310µm/s due to bubble propulsion; speeds which were comparable to those exhibited by platinum-based particles. In the case of other metal-organic framework Janus particles which were propelled by bubble ejection, the particles were prepared[97] by the selective epitaxial growth of ZIF-67 on ZIF-8. The Janus particles then catalysed the decomposition of H_2O_2 on the ZIF-67 surface but not on the zinc-containing ZIF-8 surface, thus resulting in propulsion.

Spheres are not the only choice of shape for Janus devices, and the collision of self-propelled micro-plates has been investigated[98] using Janus micro-plates which were 5 or 10µm in diameter and were driven by the catalytic decomposition of hydrogen peroxide. Periodically oscillating oxygen bubbles imparted straight or spiral motion to the plates. Because the effects of inertia are unimportant for slow motion in a low Reynolds number environment, the collisions exhibited unusual characteristics: bounce-back collision and linear collisions between the bubbles and Janus plates, as well as between 2 microplates. It was found that these 2-dimensional forms were able to destroy emergent gas bubbles. Janus disks have also been prepared[99], in which the rim was half-coated with platinum/palladium and the other half with gold. When upright with respect to the air/liquid interface they self-propelled, via bubble-growth or bubble-bursting, at a rate of about 100µm/s when submerged in peroxide; with the velocity resulting from bubble-bursting being 3 orders of magnitude higher than that resulting from bubble-growth. The distances covered were of a similar order of magnitude in both cases. The oxygen bubbles appeared due to catalysis occurring at the upper (platinum/palladium-coated) rim. The disks executed linear or rotary motions, depending upon the location of bubble nucleation. Due to the overall interfacial consumption of solute however, the two-dimensional problem of phoretic swimming is ill-posed in the sense of standard descriptions of diffusive transport where the solute concentration satisfies the Laplace equation. It becomes well-posed if account is taken of solute advection. The case of weak advection has been considered[100] and solute transport was analyzed by using matched asymptotic expansions in two separate regions. These were a near-field region in the vicinity of the swimmer, and a far-field region where solute advection approached the overall environment. An analysis was performed for a standard configuration in which half of the particle boundary was active and the other half was inert. Of most interest was the fast-reaction limit, leading to a mixed boundary-value problem in the near-field region. That problem was solved by means of conformal mapping. The asymptotic scheme led to an implicit equation for the particle velocity in the direction of the active portion of its boundary. A non-linear dependence of this velocity upon other conditions confirmed a non-vanishing effect of solute advection.

It is also possible to construct a reverse form of Janus particle which possesses an internal catalytic agent. This catalytic material, such as platinum, is embedded within a hollow mesoporous silica particle and provokes the decomposition of H_2O_2 when immersed in aqueous peroxide solution. Gaps in the non-catalytic hemisphere permit the exchange of solutes between the exterior and the interior of the particle. Upon varying the diameter of the particle, size-dependent mobility was observed[101]; with increased diffusion occurring in the case of 500nm particles and self-phoretic motion towards the non-metallic part in the case of 1.5 and 3μm particles. The direction of motion was generally explained by a theoretical model which was based upon self-phoresis. In the case of 3μm particles, a change in the morphology of the porous portion was observed which involved a change in the propulsion mechanism to bubble nucleation and ejection, together with a change in the direction of motion.

Simulations were made[102] of a spherical Janus particle undergoing exothermic surface reactions around only one pole. The model deliberately excluded self-phoretic transport. Net motion nevertheless arose from direct momentum transfer between solvent and colloid, with the speed scaling as the square root of the energy released during the reaction. Such propulsion was dominated by the short-term response of the system when neither the time dependence of the flow around the colloid, nor the solvent compressibility, could be ignored. These simulations agreed quite well with experiment.

A mesoscopic hydrodynamic model was used[103] to simulate synthetic self-propelled Janus particles which were thermophoretically or diffusiophoretically driven. A proposed model for a passive colloidal sphere reproduced the correct rotational dynamics, together with a strong phoretic effect. This colloid solution model involved a multiparticle collision dynamics description of the solvent, and combined stick boundary conditions with colloid-solvent interactions. Specific asymmetrical colloidal surfaces were introduced in order to reproduce the properties of self-phoretic Janus particles. A comparative study was made of Janus and micro-dimer phoretic swimmers in terms of their swimming velocities and induced flow behaviour. Self-phoretic micro-dimers exhibited long-range hydrodynamic interactions with a decay of $1/r^2$; similar to the decay of the gradient fields which are generated by self-phoretic particles, that can be pullers or pushers. On the other hand, Janus particles were characterized by short-range hydrodynamic interactions having a decay of $1/r^3$, and behaved like neutral swimmers.

An analysis was made of the hydrodynamic force acting on an oscillating stick-slip Janus particle, showing[104] that it could exhibit unusual history force responses that were of neither no-slip nor pure-slip type, but were mixed. It was recalled that a rigid non-slippery particle undergoing unsteady motion could experience a so-called Basset history force which exhibited a decay that was due to a Stokes boundary layer. For a uniform-slip particle with a slip-length, a persistent force plateau could replace the usual Basset decay when below the slip-stick transition point. By solving the oscillatory Stokes-flow equation, and using a matched asymptotic boundary-layer theory, it was shown that the persistent-force plateau of a uniform-slip particle could be destroyed by the stick portion of a stick-slip Janus particle. A Basset force having an amplitude smaller than the no-slip counterpart instead reappeared to dominate again the high-frequency viscous force response. This renewed Basset force, which appeared only after creating a sticking patch on a slippery particle, depended only upon the coverage of the stick-face, regardless of the slip-length of the slip face. When the stick portion was small, the renewed Basset decay included a slip plateau in its tail; thus displaying a distinctive re-entrant history force transition. If the tiny stick face was rendered slippery, the renewed Basset-decay disappeared no matter how small was the slip-length. A constant force-plateau then returned and again dominated the force response. These unusual force responses, which arose from mixed stick-slip or non-uniform slip effects, could not only provide unique hydrodynamic criteria for characterizing heterogeneous particles but also serve to manipulate and sort the particles.

A bio-catalytic Janus particle was based[105] upon mesoporous silica, with catalase being used to provoke the decomposition of peroxide so as to produce bubble propulsion, and a nickel coating being used in order to permit magnetic guidance of the particle.

Manipulation of the pH value can be used to alter the mobility of particles[106]. The gradual use of sodium hydroxide to increase the pH of a peroxide solution led to a consequent increase in the activity of particles. The addition of hydrochloric acid spoiled the structural integrity of the particles and again led to mobility changes. The marked changes in velocity of the particles could be exploited reciprocally in order to measure pH values. The methods were applied to Janus silver-based particles and to tubular bimetallic copper-platinum forms. Alteration of the pH value was a useful strategy for increasing hydrogen peroxide decomposition and enhancing oxygen-bubble propulsion.

Near-infrared laser light can be used[107] for the reversible control of a bubble-driven Janus polymer particle via illumination of the metal face of the particle at a critical concentration of peroxide.

Iridium-based bubble-propelled graphene Janus particles having a very high surface area, but very low catalyst content, were investigated[108], showing that just 0.54at% of iridium catalyst provided rapid motion of graphene having a specific surface area of 316.2m^2/g. The particles could be easily prepared via the thermal exfoliation in hydrogen of an iridium-doped graphite oxide precursor composite. The propulsion was provided by the decomposition of hydrogen peroxide at the iridium-catalyst locations.

Silver-silica and silver-polystyrene particles were propelled[109] as usual by the splitting of peroxide and the generation of oxygen bubbles. The bubble-particle combinations were stable, and could be manipulated by altering hydrophilic-hydrophobic interactions with the surface. A transition between 2-dimensional and 3-dimensional motion could be arranged by changing the hydrophobicity of the substrate.

Silver-based Janus nanoparticles are of interest in many applications because of their excellent photoelectric performance, but it is difficult to prepare this type of Janus nanoparticle at a large scale and with high uniformity. A simple wet-chemical strategy was presented[110] for the synthesis of Ag-MoS$_2$ Janus nanoparticles by adding a suitable amount of AgNO$_3$ to a 200nm MoS$_2$ nanosphere-containing aqueous solution. These nanoparticles constituted a new type of heterogeneous nanostructure with each particle comprising one minute silver-dot decorated MoS$_2$ nanosphere and one polyhedral silver single crystal which was about 100nm in size. Joined together as a dimer, they resembled a snowman. Such Janus nanoparticles or dimers could not be detached even by high-power ultrasonic vibrations, thus demonstrating their strong adhesion. The morphology and size of the polyhedral silver nanocrystals attached to MoS$_2$ nanospheres could moreover be controlled by the amount of added AgNO$_3$, the reaction time and the temperature. A growth model for the formation of these nanoparticles was based upon aggregation of the minute silver dots and an oriented connection mechanism. These Ag-MoS$_2$ nanoparticles could be used as nanomotors due to their asymmetrical catalytic structure and resultant self-propulsion in H$_2$O$_2$ solution. They exhibited good directional motion, with a marked dependence of their speed upon the peroxide concentration.

So-called coconut particles, involving platinum and the partial or complete etching of silica, were investigated[111]. Although the inner and outer surfaces were both made of platinum, motion occurred because the convex surface was able to generate oxygen bubbles. This demonstrated that both chemical asymmetry and geometrical asymmetry could drive rapid propulsion. A much higher velocity was achieved by partially etched coconut particles as compared with the velocities of Janus or fully-etched shell-like particles.

Janus particles need not depend upon a single fuel such as hydrogen peroxide. An example has been described[112] in which energy was obtained from 3 fuels. Aluminium-palladium particles were prepared by depositing a palladium layer onto one side of aluminium microspheres. These were propelled by hydrogen bubbles which were generated by the reaction of aluminium when in strongly acidic or alkaline environments (figure 15). They could also be propelled by oxygen bubbles which were produced by the partial palladium coatings when placed in hydrogen peroxide solutions. Velocities and lifetimes of 200µm/s and 480s were observed for strongly alkaline and acidic media, respectively. The particles could adapt to the presence of a different fuel without affecting the propulsion behavior.

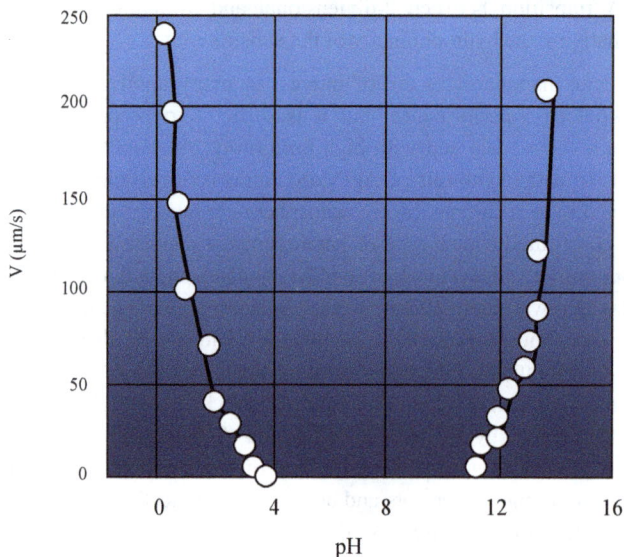

Figure 15.. Velocity of 20µm aluminium-palladium particles at blood heat (37C) as a function of pH

At the other extreme, entirely water-driven Janus particles were described[113] which consisted of partially-coated aluminium-gallium microspheres which were prepared by microcontact-mixing of aluminium microparticles and liquid gallium. Here the ejection of bubbles of hydrogen from the alloy hemisphere, due to reaction with the water, again provided propulsion. This was because the presence of gallium impeded the protective effect of the aluminium's oxide, without which the notoriously reactive bare aluminium would behave like sodium in water. It was therefore not surprising that these particles could move at 3mm/s (150 body-lengths/s) and exert forces greater than 500pN.

A Janus polyelectrolyte multilayer hollow particle was developed[114] which could transport materiel. It consisted of partially-coated dendritic platinum nanoparticles. The platinum on one side naturally decomposed hydrogen peroxide and the resultant oxygen bubbles provided propulsion at more than 1mm/s (125 body-lengths/s) and exerted forces greater than 75pN. The particles could also be guided by using an external magnetic field.

A simple approach, based upon bipolar electrochemistry, was demonstrated[115] which permitted the bulk preparation of carbon microtubes that sported a platinum cluster at one end. Its presence catalytically decomposed hydrogen peroxide and, as usual, led the resultant oxygen bubbles to propel the microtube. The motion which was produced varied from linear to circular, depending upon the position of the platinum cluster.

Non-conductive Janus particles, which had been made by coating fluorescent polymer beads with platinum on one hemisphere, were monitored by means of video microscopy[116]. This showed that they were propelled, in hydrogen peroxide solution, away from the platinum coating. Because the latter shielded the fluorescence signal from half of each particle, it permitted the orientation to be observed directly and related to the resultant direction of motion. This direction was found to support the operation of both bubble-release and diffusiophoretic propulsion mechanisms. Optical microscopy, combined with imaging software, has commonly been used to study the particles. An alternative is the use of particle-electrode impact voltammetry. Disturbance of the diffusion layer at the electrode interface, resulting from motion of a particle in solution can even result in spikes resulting from electrochemical signals[117]. This has been demonstrated in the case of silver Janus particles and tubular copper-platinum particles.

A thermosensitive Janus particle was synthesized[118] by using a combination of vacuum sputtering and chemical polymerization. The micromotor involved catalytically active platinum and a thermosensitive shell which affected movement by altering the system temperature. When the movement of the autonomous micromotor was observed *in situ* by means of hot-stage and cold-stage optical microscopy, it was obvious that the particles moved in one direction when H_2O_2 was added, but their velocity became much lower when the temperature was increased from 25 to 40C in 10%H_2O_2 aqueous solution.

External propulsion

radiation

As well as being propelled by reactions with the medium (fuel) in which they are immersed, Janus particles can also be powered by external means although it is often moot whether the external stimulus is merely driving one of the self-propulsion mechanisms rather than acting directly on the particle. For example, in the case of particles which comprise both photocatalytic TiO_2 and catalytic platinum surfaces[119], the chemical reactions on the differing surfaces can be tailored by using chemical and radiation effects to generate counteracting propulsive forces on the catalytic and photocatalytic sides. Variation of the surface chemistry of such a particle leads to reversals of the direction of motion which reflect the state of the local ion concentration and thus the predominant propulsive force. A gold layer below the platinum surface can play an important role in determining the propulsion mechanism. Such an optical braking system permits control of the chemical propulsion via photocatalytic reaction on the TiO_2 which balances the chemical propulsive force generated on the platinum side. The advantages of light-driven propulsion include remote control, in that the velocity can be varied or the particle stopped and started at will. The particles can be in the form of tubes and spheres, or have more irregular shapes. The activation mechanisms include photocatalysis, photothermal effects, photo-isomerization and photo-induced deformation. The most important feature of the propulsion mechanism is the gradient field which exists around the particle. This can be solutal, electrical, thermal or radiative. An asymmetrical particle structure or light field is the key factor in particle propulsion. Another ambivalent case, in which it is unclear whether the particle is self-propelled or propelled from without, is the steering of Janus-type particles[120]. Real-time information on the particle position and orientation can be used to change optically the self-propulsion mechanism of the particle. Its orientational Brownian motion provides a reorientation mechanism. A particle-size dependence of the photophoretic propulsion velocity indicates that photon-nudging provides increased positional accuracy with decreasing particle radius. In so-called photon-nudging, a laser can intermittently push a particle along its body-axis via a combination of radiation pressure and photophoresis[121]. As rotational random walks reorient the particle micro-swimmer, propulsion is provided only when the particle is aligned with a given target, thus offering control. A Langevin-type equation of motion can be derived which combines these concepts so as to describe the dynamics of a stochastically controlled particle. One experimental test of the equation has revealed that a temperature differential of about 7K on the particle surface could generate a photophoretic propulsive force of about 0.1pN. It was also noted that there were 2 useful parameters which could inform the manipulation of particles: one was the number of random-walk turns which the particle exhibited, and the other was the photon-nudge distance within the rotational diffusion time.

Janus particles in the form of gold-capped colloidal spheres in a critical binary liquid mixture perform Brownian motion when illuminated by light. When investigated more closely, it was found[122] that illumination-induced heating caused local asymmetrical de-mixing of the binary mixture, thus generating a spatial chemical concentration gradient which was then responsible for the self-diffusiophoretic motion. This effect was studied as an aspect of the functionalization of the gold cap, the particle size and the intensity of illumination. The functionalization governed which component of the binary mixture was preferentially adsorbed at the cap, together with the direction of motion; towards or away from the cap. The particle size governed rotational diffusion, and thus the random reorientation of the particle. The intensity meanwhile governed the degree of heating and therefore the extent of motion. The dependence of the swimming strength upon the illumination intensity was exploited so as to investigate the behaviour of a micro-swimmer in a spatial light gradient, where the swimming properties were space-dependent.

It was initially reported[123] that gold-capped Janus particles, immersed in a near-critical binary mixture, could be propelled by illumination. A non-isothermal diffuse-interface analysis was made of the self-propulsion mechanism of a single colloid, and motion was attributed to body-forces at the edges of a micron-sized droplet which nucleated around the particle. The concept of a surface velocity could not account for the self-propulsion. The particle-velocity was related to the droplet shape and size, and was governed by a critical isotherm. Two distinct propulsion-regimes existed; depending upon whether the droplet partially, or completely, covered the particle. The dependence of the velocity upon temperature was non-monotonic in both regimes. Fuel-free near-infrared driven Janus particles were created[124] via template-assisted polyelectrolyte layer-by-layer assembly, and a gold layer was then sprayed onto one side. These particles could attain a maximum speed of 42μm/s in water, and their movement was attributed to self-thermophoresis. That is, the asymmetrical distribution of gold generated a local thermal gradient which in turn generated a thermophoretic force that propelled the particle. By introducing a tailored optical asymmetry into a particle it was possible to guide it by controlling the light-frequency, without regard to the direction or shape of the light beam. Stochastic simulations demonstrated[125] the guidance of a 2-faced nanoparticle in which optically-induced thermophoretic drift was the propulsion mechanism. By taking advantage of the difference in resonant absorption spectra of the 2 materials, it was possible to create a bi-directional local thermal gradient that could be controlled externally. Unlike shape or coherence, the frequency of the light beam was maintained

Materials Research Forum LLC
https://doi.org/10.21741/9781644901199

even in strongly scattering media. Examination of the mesoscopic collective motion of self-propelling gold-silica microspheres under laser irradiation, by means of long-term particle-tracking, showed[126] that a suspension could be driven from equilibrium to near-equilibrium and eventually to far-from-equilibrium by adjusting the excitation laser intensity. At low laser intensities, the suspension was driven to a near-equilibrium state via homogeneous superdiffusive motion; the strength of the increased superdiffusion being monotonically related to the laser intensity. At high laser intensities, motility-induced phase separation occurred leading to the coexistence of dense clusters and very few individual particles. This led on to highly heterogeneous dynamics, with fewer jammed mobile clusters and fast-moving particles, that finally suppressed any increased superdiffusion.

A study of ZnO-platinum Janus particles with atomically smooth interfaces revealed rapid (15 body-lengths/s) light-driven fuel-free propulsion[127]. The velocity was further increased, by some 60%, following the addition of a few atomic interlayers of amorphous TiO_2 photocatalyst. These new ZnO/TiO_2 photocatalytic interfaces provided type-II heterojunctions, led to an increase in the number of electron/hole pairs and improved electron-transfer to the platinum. This effective charge separation/transfer resulted in more rapid electrophoretic motion of the ternary ZnO/TiO_2-platinum particles. Further investigation of these mesoporous ZnO-platinum Janus particles was based[128] upon 2 models for ZnO semiconducting particles having distinct surface morphologies and pore structures. These were prepared by the self-aggregation of primary nanoparticles and nanosheets into nanoscale rough and smooth microparticles, respectively. The self-assembled nanosheet model for smooth particles provided a large surface area for the formation of a strongly adherent continuous platinum layer, whereas discontinuous platinum occurred within inter-nanoparticle pores in the self-assembled nanoparticle model.

Bismuth oxyiodide based Janus particles have been studied[129] which could be activated by blue and green light. They were propelled by photocatalytic reactions in pure water under visible light, without requiring any additional chemical fuel. Remote control was possible by modulating the intensity of the visible light. The self-electrophoresis mechanism was confirmed by noting the effects of layers of Al_2O_3, platinum and gold upon the velocity.

The collective behavior of visible-light photochemically-driven plasmonic Ag-AgCl Janus particles, when surrounded by passive polystyrene beads, was studied[130]. The movement of single Janus particles and of small (2- or 3-particle) and large (more than 10-particle) clusters interacting with the passive beads was analyzed. The movement of active particles could be regulated by adjusting the number of single Janus particles in the

Materials Research Forum LLC
https://doi.org/10.21741/9781644901199

cluster. The Langevin equations of motion for self-propelled Janus particles and migrating passive polystyrene beads were solved numerically by using molecular dynamics simulations while considering core-core interactions, short-range attraction and an effective repulsion due to light-induced chemical reactions.

In order to determine how various metals affect the velocities of metal-TiO_2 particles under ultra-violet irradiation in pure water, samples of platinum-TiO_2, copper-TiO_2, iron-TiO_2, silver-TiO_2 and gold-TiO_2 Janus particles were prepared[131]. They were chosen to have differing chemical potentials, and catalytic effects upon water-splitting, given that these properties were expected to alter the photo-electrochemical response and propulsion rate. It was deduced that the effective velocity was a result of the synergistic interaction of chemical potential and catalysis. The platinum-TiO_2 Janus particles were the fastest (figure 16), but ultra-violet exposure tended to make these, and the other particles, form chains which then affected the velocity.

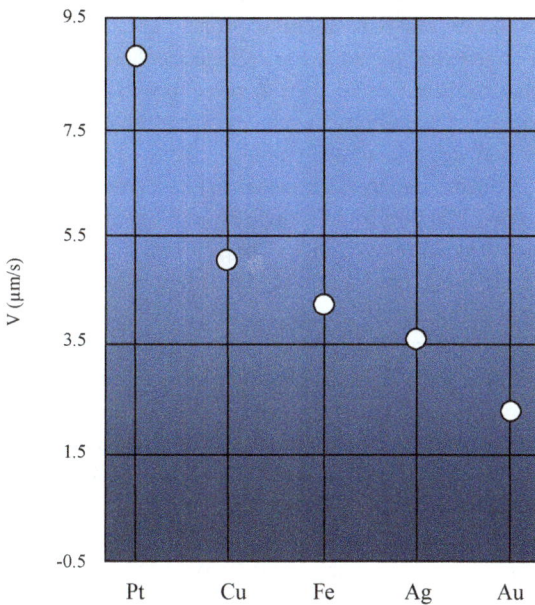

Figure 16. Average velocities of various metal-TiO_2 particles under ultra-violet light

Simple self-propelled spheres can assemble into more complex structures which exhibit a rotational motion that is perhaps associated with translational motion, as in flagella. A combination of induced dipolar interactions and a bonding step was used to create permanent linear bead-chains which comprised self-propelling Janus spheres having a well-controlled internal structure. A study was made[132] of how flexibility between individual swimmers in a chain could affect the swimming behaviour. Permanent rigid chains exhibited only rotational or spinning motions, while longer semi-flexible chains exhibited both translational and rotational motions which resembled flagella-like motion. It was possible to reproduce experimental results by performing numerical calculations using a model which included full hydrodynamic interaction with the fluid. This method was general and might be used to design self-propelled colloids which exhibited complex swimming behaviours by using various complex building blocks and associated linkage-flexibility.

Hollow mesoporous TiO_2-gold Janus particles were prepared[133] which exhibited increased velocities under low-intensity ultra-violet light; in both the presence and absence of H_2O_2. The particles moved due to self-electrophoresis, when ultra-violet light was played on them, and there was a 3-fold increase in velocity as compared with solid Janus TiO_2-gold particles. The increase in velocity was attributed to the increased surface area of the porous particles and to their hollow nature. The particles could also be rendered steerable by incorporating a thin cobalt layer. Particles offering photocatalytic propulsion and directional control could also be based[134] upon less expensive iron, while retaining the traditional photocatalyst, TiO_2. The velocity still increased with increasing fuel concentration and light intensity and the travel direction could moreover be controlled by using an external magnetic field. A light-driven $1.0\mu m$-diameter photocatalytic TiO_2-gold Janus particle, having wireless steering and velocity control, was described[135]. It could be powered in pure water by using an ultra-violet light intensity of only $2.5 \times 10^{-3} W/cm^2$. At an intensity of $40 \times 10^{-3} W/cm^2$, the particle could attain a velocity of 25 body-length/s. This was comparable to the velocities of platinum-based chemically-induced self-electrophoretic Janus particles. The photocatalytic propulsion process could be stopped and started by changing the incident light modulation. The velocity of a photocatalytic TiO_2-gold Janus particle could be further increased by increasing the light intensity or by adding just 0.1% of H_2O_2.

It is indeed often the case that the so-called external propulsion of a particle nevertheless requires the intervention of the particle's environment. In the case of photocatalytic TiO_2-platinum Janus particles, the effects of structural parameters and fuel concentration were determined[136]. Experimental data showed that crystallinity enhancement, loading with up to 1% of platinum nanoparticles, coating with less than 4nm of platinum and increasing the fuel concentration all increased the light-controlled velocity and driving force of the particle. Upon optimizing its structure and fuel concentration, the particle velocity and driving force of the Janus particle could attain 30μm/s and 0.341pN, respectively. The motion of water-fuelled TiO_2-platinum Janus particles was governed[137] by light-induced self-electrophoresis under local electrical fields which were generated by asymmetrical water oxidation and reduction on its surface. These particles could interact with each other via light-switchable electrostatic forces: continuous and pulsed ultra-violet irradiation could make the particles aggregate or separate, respectively. Because of the increased mass-exchange which occurred between the environment and active particles, separated particles which were powered by pulsed ultra-violet irradiation exhibited a much higher activity for photocatalytic degradation of organic dye than did aggregated particles. In related work[138], a bubble-propelled photo-activated amorphous TiO_2-gold Janus particle was created which exploited photocatalytic H_2O_2 decomposition over *in situ* H_2O_2-sensitized amorphous TiO_2 under ultra-violet radiation. The quantum efficiency of the oxygen-bubble evolution arising from the photocatalytic decomposition of H_2O_2 attained 28% and the power-conversion efficiency attained 1.28×10^{-9}. This led to a maximum speed of 135μm/s. The motion and velocity of the particle could be remotely controlled, within less than 0.1s, by changing the intensity of the ultra-violet radiation. It was noted[139] that the aggregation of passive silica colloids into well-controlled 2-dimensional assemblies was controlled by a small number of self-propelled active colloids. These were titania–silica Janus particles that were propelled when illuminated using ultra-violet light. The strength of the interaction and the extent of the clusters could be regulated by varying the light intensity. A relatively small number of active colloids was required in order to trigger assembly.

The dye-enhanced self-electrophoretic propulsion of light-driven TiO_2-gold Janus particles has been observed[140] in aqueous solutions. As compared to their velocity in pure water, enhancement factors of 1.7, 1.5 and 1.4 were observed in aqueous solutions of methyl blue (10^{-5}g/l), cresol red (10^{-4}g/l) and methyl orange (10^{-4}g/l), respectively. The velocity-change depended upon the type of dye, due to variations in their photo-degradation rates. When a nickel layer was inserted between the gold and TiO_2 layers, the particle could be steered by using an external magnetic field. A light-driven gold-WO_3-C Janus particle, based upon colloidal-carbon WO_3-nanoparticle composite spheres, was prepared[141] by using a single-step hydrothermal method. It could move at 16μm/s in aqueous media, under 40mW/cm^2 ultra-violet illumination, due to diffusiophoresis. The propulsion of such 1.0μm Janus particles could be provoked in pure water by ultra-violet light, without requiring any external chemical fuel. By depositing a nickel layer between the gold and the WO_3, the particle could be controlled by an external magnetic field. The

Janus particles exhibited a high sensitivity to extremely low concentrations of sodium-2,6-dichloroindophenol and Rhodamine-B, and their velocity could be increased to 26 and 29μm/s in aqueous solutions of 5 x 10^{-4}wt% sodium-2,6-dichloroindophenol and 5 x 10^{-7}wt% Rhodamine-B, respectively. This was attributed to an enhanced diffusiophoretic effect that resulted from the rapid photocatalytic degradation of the sodium-2,6-dichloroindophenol and Rhodamine-B by WO_3.

The behaviour of Janus particles that self-propel in solution due to light-activated catalytic decomposition of the surrounding fluid has recently been analyzed[142] anew by using a model, for a photo-active self-phoretic particle, that took account of the so-called self-shadowing of the illumination by the opaque catalytic side of the particle. This self-shadowing could cause rotation of the catalytic cap toward the light-source (phototaxis) or have the opposite effect (anti-phototaxis), depending upon the details of the particle properties. Upon including the effect of thermal noise, it could be shown that the distribution of particle orientations could be described as a Boltzmann distribution involving a non-equilibrium effective potential. The average vertical velocity of a phototactic particle exhibited a super-linear dependence upon the light intensity, while that of an anti-phototactic particle exhibited a sub-linear dependence upon the intensity. Inorganic particles have been based[143] upon various propulsion mechanisms, but the only means for controlling their direction was to incorporate a ferromagnetic material and to use an external magnetic field for steering. A particle was prepared which could sense the direction of illumination and orient itself towards the external source. It consisted of a Janus nanotree which contained a nanostructured photo-cathode and photo-anode, at opposite ends, that released cations and anions respectively, and propelled the particle by self-electrophoresis. Chemical modification could affect the zeta potential of the photo-anode and cause the particle to exhibit positive or negative phototaxis.

Unlike the common peroxide-fuelled particles, glucose-fuelled ones are propelled by enzyme mechanisms, which tend to impart lower velocities. New glucose-fuelled, Cu_2O plus N-doped carbon nanotube particles, can be activated[144] by visible-light photocatalysis and their velocities can attain 18.71μm/s. Such a performance is comparable to that of the platinum-based catalytic Janus particles which are usually powered by relatively toxic peroxide fuel. The velocities of the new particles could be controlled by changing the carbon-nanotube content, the glucose concentration or the light intensity. The particles also exhibited negative phototaxis, in that they moved away from the light. The enzyme activity, and therefore the velocity, could also be varied by changing the substrate concentration or by adding inhibitors[145]. Nanoscale swimmers are more affected by Brownian fluctuations than are micro-sized particles and therefore their motion is better presented as an increased diffusion, with respect to the passive case.

Micro-particles could however overcome the fluctuations and exhibit propulsive or ballistic trajectories. The principle that symmetry constrains dynamics was tested[146], experimentally and by molecular dynamics simulation, for a hot Janus swimmer operating far from thermal equilibrium. This established scalar and vectorial steady-state fluctuation theorems plus a thermodynamic uncertainty relationship which linked the fluctuating particle current to its entropy production at an effective temperature. A Markovian model clarified the underlying non-equilibrium physics.

Particles which were based upon either polystyrene, or upon polystyrene coated with a rough silica shell, were studied[147]. Directional propulsion at higher speeds was found for the polystyrene-SiO_2 particles than for the polystyrene ones. Optical reconstruction microscopy was used to detect single urease molecules which were conjugated to the particle surface. An asymmetrical distribution of enzymes around the particle surface was observed for both particles. This indicated that the enzyme distribution was not the sole parameter affecting motion. There was a 10-fold increase in the number of urease molecules for polystyrene-SiO_2 particles as compared with polystyrene ones. Both speed and force depended upon the enzyme coverage in a non-linear manner. In order to break symmetry for active propulsion, it was found that a threshold number of enzyme molecules was required. Sub-micron Janus particles have been prepared[148] which had one hemisphere decorated with the enzyme-pair glucose oxidase and catalase and which used glucose as fuel. The colloids exhibited a glucose concentration-dependent enhanced diffusion behavior. Two further innovations[149] coupled an enzyme such as glucose oxidase to an inorganic nanoparticle such as platinum in order to provide power, or used a peptide-fuelled trypsin motor. Both of them increased the diffusive properties of Janus particles by using harmless biologically available fuel molecules. An improvement in a particle's properties resulted from using both innovations simultaneously. The incorporation of magnetic nanoparticles further allowed the particle to move in a magnetic gradient upon applying an external magnetic field. An enzyme-powered Janus particle was prepared[150] by half-coating a 4nm layer of silicon dioxide onto a 90nm mesoporous silica nanoparticle, thus imparting an asymmetry. The particle was conventionally powered by H_2O_2 decomposition which was triggered by catalase enzyme, spread onto one face of the nanoparticle. The apparent diffusion coefficient was increased by 150%, as compared with Brownian motion, at a peroxide concentration of less than 3wt%. A further particle design[151], based upon hollow mesoporous silica, was powered by the biocatalytic decomposition of urea at physiological concentrations. Directional self-propelled motion persisted for more than 600s at an average velocity of up to 5 body-lengths/s, and the velocity could be controlled by chemically inhibiting or re-activating the enzymatic activity of urease. The incorporation of magnetic material again permitted remote control of the movement direction. The mesoporous structure could also accommodate molecules and carry them around. In a further development, such particles were powered[152] by the biocatalytic reactions of catalase, urease or glucose oxidase. The apparent diffusion coefficient was increased by up to 83%. Optical tweezers revealed that a force of 64fN was required in order to counteract the self-propulsion force that was generated by a single particle. Enzyme-immobilized particles having asymmetrical Janus or symmetrical enzyme distributions were used to investigate[153] the interaction between enzyme reaction and motion, using lipase and urease as examples. All of the enzyme-immobilized particles exhibited substrate-concentration dependent diffusion augmentation, but the Janus-type particles underwent the greatest (60%) increase in self-propulsion. Enzymatic reactions were aided more by the motion of the Janus enzyme-immobilized particles. Self-propelled cyclic motion of Janus particles has been observed[154] in static line optical tweezers. The particle was a 5μm-diameter polystyrene sphere which was half-coated with a 3nm gold film. The cyclic translational and rotational motion was attributed to the combined action of the gold face's orientation-dependent optical propulsion force, the gradient optical force and a symmetry-breaking induced optical torque in various regions of the optical tweezers. This constitutes a means for remotely controlling the motion of microscopic particles by using a static optical field having a suitably designed non-uniform intensity profile. It had previously been reported[155] that a Janus particle which consisted of a silica sphere having a hemispherical gold surface could be stably trapped by optical tweezers, and moved in a controlled manner along the axis of a laser beam. This behaviour was attributed to the interplay of optical and thermal forces. That is, scattering forces oriented the asymmetrical particle while strong absorption on the metallic side produced a thermal gradient which drove its motion. An increase in laser power induced an upward movement of the particle, while a decrease induced downward movement. When a spherical gold nanoparticle was trapped at the same time, the distance between the 2 particles could be controlled.

It has been shown[156] that Janus particles can be oriented by an homogeneous thermal field. Multiscale simulations and theory indicated that the internal mass gradient of the particle could increase or reverse the relative orientation of the particle with respect to the thermal field. This was due to a coupling of the internal anisotropy of the particle and the heat flux. This helped to explain previous experimental observations.

A new propulsion method for light-driven self-propelled porous Janus colloid particles has recently been identified[157]. The mechanism is based upon the occurrence of light-driven diffusio-osmotic flow around the particle. Photosensitivity is imparted to the particle by means of a surfactant-containing cationic azobenzene. This undergoes photo-isomerization from a hydrophobic *trans*-state to a hydrophilic *cis*-state under illumination. When negatively charged porous silica particles are immersed in a suitable aqueous solution, they absorb molecules when in the *trans*-state but reject them when in the *cis*-state. When blue light provokes simultaneous *trans-cis* and *cis-trans* isomerization at the same time, the particle moves due to the outflow of *cis*-isomers and the in-flow of *trans*-isomers. This occurs because a hemispherical metal cap partially breaks the symmetry of an otherwise radially directed local flow around the particle, leading to self-propulsion. These particles exhibited super-diffusive motion at some 0.5μm/s, with a persistence length of about 50μm. When confined to microchannels, the direction was stable for up to 300μm before rotational diffusion changed it. Particles which were formed into dimers of various shapes could even follow circular trajectories. The strength of self-propulsion could be controlled by changing the intensity and wavelength of the irradiation.

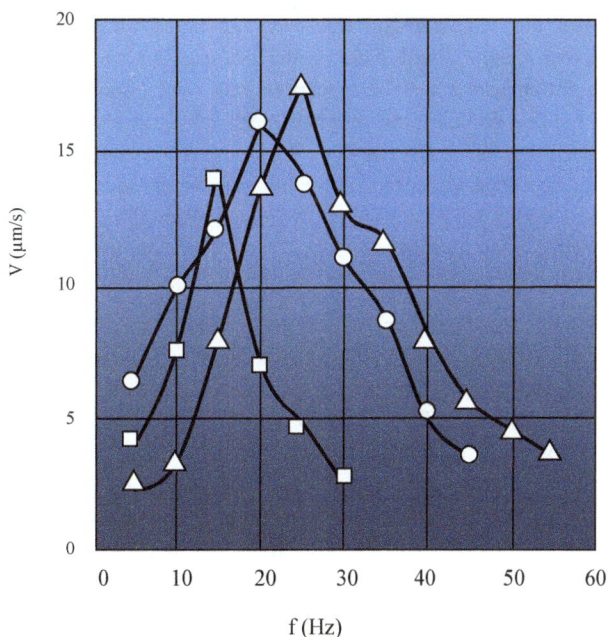

Figure 17. Velocity of nickel-SiO₂ surface particles as a function of magnetic field frequency. Triangles: 2.5μm, circles: 3μm, squares: 5μm particles

The diffusio-osmotic slip propels a particle even in the absence of an externally applied concentration gradient. A study has been made[158] of the effects of viscoelasticity and shear-thinning upon diffusio-osmotic slip on an active surface. These were described as being second-order fluid and Carreau models, respectively. By using matched asymptotic expansions, an analytical expression was found for the slip modification which was introduced by the non-Newtonian behaviour. This revealed that the modification of slip velocity which arose from polymer elasticity was proportional to the second tangential derivative of the concentration field. Using the reciprocal theorem, an estimate was made of the effect of this modification upon the swimming velocity of a Janus sphere. In the case of a second-order fluid, the contribution was non-negligible and its sign depended upon the surface coverage of the activity. In the case of a Carreau fluid, the contribution was more marked and always increased the swimming velocity.

An experimental study, using Janus colloids, of the sedimentation of dilute active colloids confirmed[159] the existence of an exponential density profile. There existed a polarized steady state when the sedimentation rate was not much lower than the propulsion velocity. The experimental distribution of the polarization was closely predicted by theory, with no fitting parameter. Three expressions were proposed for measuring the pressure of sedimenting particles: the weight of particles above a given height, the flux of momentum and active impulse and the force density as recorded by pressure gauges. The behaviour of Janus particles under free-fall conditions, where sedimentation effects could not interfere, has recently been determined[160]. This revealed the 3-dimensional dynamics of suspensions when a strong homogeneous light source was used to provoke the self-propulsion of Janus particles.

Two-dimensional particles based upon carbon nitride can be very efficient light-driven microswimmers in aqueous media[161]. Due to the excellent photocatalytic properties of poly(heptazine imide), microswimmers can be activated by using either visible or ultra-violet light, plus various capping materials (gold, platinum, silica) and fuels (H_2O_2, alcohol). Diffusiophoretic propulsion is the predominant mechanism, and oxygen-reduction reactions are the main feature under ambient conditions for metal-capped particles. Hydrogen evolution reactions are the usual cause of propulsion under ambient conditions when using alcohols as fuels. By exploiting the intrinsic solar energy storage ability of poly(heptazine imide), photocapacitive Janus microswimmers could be charged using solar energy. This then permitted continued light-induced propulsion, even in the absence of illumination, for at least 0.5h. This implied the possible extension of drug delivery routes.

Based upon experiments in which Janus particles were made into active swimmers by illuminating them with laser light the effect of applying a light pattern to the sample, thus creating activity-inducing zones or active patches, was studied[162]. A system of interacting Brownian diffusers was simulated which became active swimmers when moving within an active patch. To some degree, the effect of spatially inhomogeneous activity was qualitatively similar to that of a temperature gradient. In the case of asymmetrical patches, this analogy failed.

magnetic field

The motion of spherical 5μm platinum-SiO_2 Janus particles in 2-dimensional space under the control of a magnetic field was studied[163]. The particles were propelled with sufficient force to permit them to overcome drag and gravitation and to move downwards and upwards. The particles moved downwards and upwards at average speeds of 19.1 and 9.8μm/s, respectively. A closed-loop control system permitted localization in 3-dimensional space to within an average region-of-convergence of 6.3μm. Harking back to the biological inspiration for Janus particles, a magnetic nickel-SiO_2 surface particle was studied[164] that could attain a velocity of up to 18.6μm/s (4 body-lengths/s) in an oscillating magnetic field of 25Hz and 2.7mT (figures 17 and 18). Two of them could form a micro-dimer by magnetic dipolar interactions. The oscillating magnetic field could also accurately steer these surface particles through complicated structures.

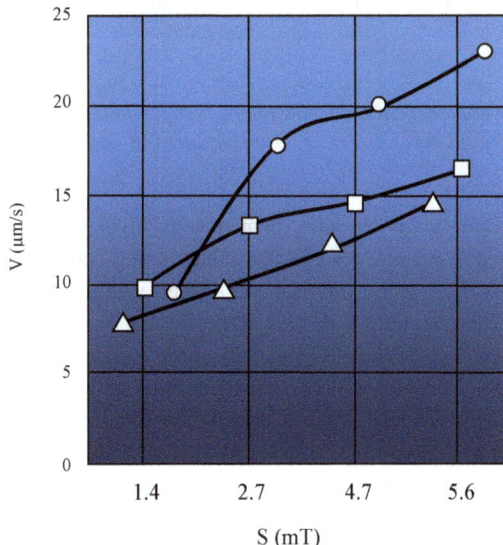

Figure 18. Velocity of nickel-SiO_2 surface particles as a function of magnetic field strength. Circles: 24Hz, squares: 18Hz, triangles: 12Hz

Materials Research Forum LLC

https://doi.org/10.21741/9781644901199

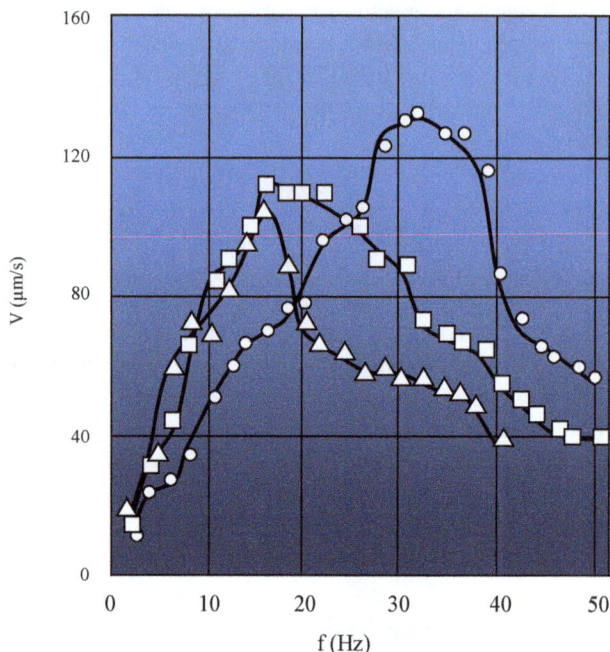

Figure 19. Velocity of nickel-SiO₂ micro-dimer particles in a magnetic field, as a function of the driving frequency. Circles: 5+5µm, squares: 8+8µm, triangles: 10+10µm

Hemispherical polymer Janus particles, coated with silver or nickel on the equatorial plane, were dispersed in water and subjected to alternating-current electric and stationary magnetic fields[165]. They were oriented with the equatorial plane parallel to the alternating-current electric field due to electric-field induced dipole orientation. The nickel-coated particles could self-assemble to form chain-like structures which were aligned in the direction of the stationary magnetic field. When alternating electric fields and stationary magnetic fields were simultaneously applied, nickel-coated hemispherical particles alone oriented themselves so that the equatorial plane was parallel to both the electric and magnetic fields. Given that the particle was magnetized in the plane, its direction reversed when the magnetic field was reversed. Such effects were possibly relevant to the behaviour of moving particles. Homogeneous magnetic fields cannot exert a net force on a colloidal particle but, when combined with the particle's position on a

Materials Research Forum LLC
https://doi.org/10.21741/9781644901199

curved interface, even static homogeneous fields can produce rapid particle motion. The effect was demonstrated[166] by using magnetic Janus particles, having an amphiphilic surface, which were adsorbed at the spherical interface of a water drop in decane. The imposition of a static homogeneous field produced movement to the drop equator, where the particle's magnetic moment could align itself parallel to the field. The effective magnetic force on the particle was linearly proportional to the curvature of the interface. A uniform rotating magnetic field has been used[167] to propel Janus micro-dimer particles which consisted of nickel-SiO_2 Janus microspheres that were held together magnetically. Their velocity could be varied by adjusting the magnetic field frequency (figures 19 and 20), and could attain a maximum of 133µm/s (13.3 body-lengths/s) at a frequency of 32Hz. The particles could also be guided through complex environments, and near to obstacles.

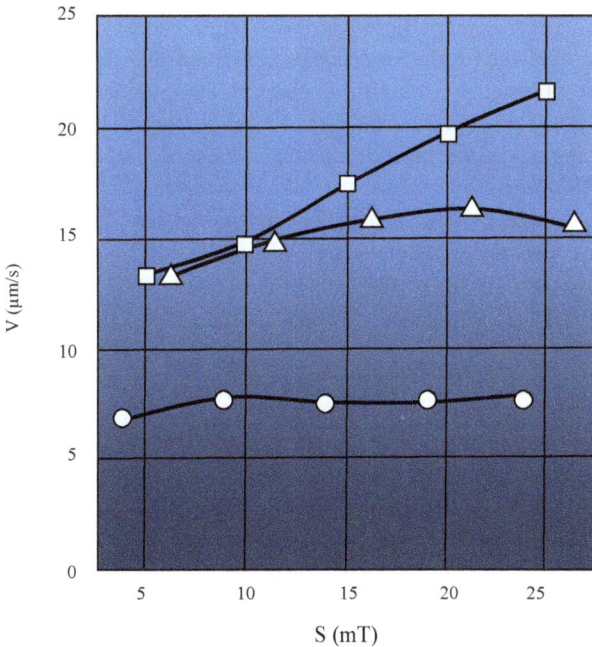

Figure 20. Velocity of nickel-SiO_2 micro-dimer particles in a magnetic field, as a function of the magnetic field strength at a driving frequency of 1Hz. Squares: 8+8µm, triangles: 10+10µm, circles: 5+5µm

A technique was presented[168] for the near-surface control of magnetic nanorod swimmers and for micromanipulation. The rotation of magnetic Janus nanorods showed that this movement, when performed near to a plane, resulted in a predictable translational motion. The nanorod plane-of-rotation was almost parallel to the plane, with the angle between the tilted rod and the plane being between 0° and 20°. Orthogonal magnetic fields could arbitrarily control the in-plane motion. A proposed model for translation incorporated symmetry-breaking via increased drag at the no-slip surface boundary. This method offered a considerable ability to steer the rod. Janus nanorods could be attached to the cell surfaces and used to manipulate individual cells. This process could also be used to manoeuvre considerable payloads. Polystyrene micro-beads could be captured and manipulated.

electric field

Although not relevant to propulsion it is interesting to note that, when subjected to a rotating electric field, the rotation direction of a platinum-SiO_2 Janus particle was mainly opposite to the direction of the electric field; similar to the behaviour of metallic particles. This rotation direction could reverse at frequencies below 1kHz or above 1MHz; depending upon the metallic coating thickness and the conductivity of the solution. The electrorotation attained its maximum angular velocity at a characteristic frequency which increased with the metallic coating thickness[169]. It could be an order-of-magnitude higher than that of an entirely metal-coated particle. It was suggested that the electrorotation response of the particle exhibited dielectric and metallic features and that the responses were not simply the averaged responses of its two sides. When Janus colloidal particles were subjected[170] to an alternating-current electric field, the velocity and direction of propulsion and the strength of the attractive interaction between particles could be controlled by varying the frequency of the applied field and the ion concentration of the solution. A strong attractive force which appeared at high ion concentrations gave rise to the formation of chains of particles. This was explained in terms of the quadrupolar charge distribution on the particles. Such chain formation occurred regardless of the direction of propulsion. The beat frequency of the chains depended upon the applied voltage and upon the propulsive force. The relationship between the beat frequency and the propulsive force deviated from theoretical predictions but this was attributed to the attractive interaction being mediated by the quadrupolar distribution of the induced charges. It has been suggested[171] that liquid droplets could serve as Janus particles instead of solids, and this is discussed elsewhere in the present work, with regard to phoretic self-propulsion. The self-propulsion of a Janus droplet in a solution of surfactant which reacts with only one half of the droplet surface has been studied theoretically[172]. The droplet acts as a catalytic motor and creates a concentration gradient which produces motion driven by surface tension. The self-propulsion velocity is relatively high, and such a catalytic motor has the advantages of simple preparation and inherently neutral buoyancy. Unlike a single-fluid droplet, which may undergo self-propulsion due to a symmetry breaking instability, the Janus droplet has no stability threshold and the droplet radius can be arbitrarily scaled down.

It was demonstrated that liquid droplets could be used to build an active propulsion system, with the droplet system being easily variable from a core–shell to a Janus arrangement by using surfactant-based methods and many micro-fluidic methods exist for preparing suitable droplets[173,174,175,176,177]. There had already been success in using the electro-hydrodynamic flow produced by an alternating electric field to induce the motion of colloidal particles, and its utility in propelling ionic liquids was shown. Returning to solid Janus particles, the details of their behaviour can be studied[178] in detail by using an interdigitated microelectrode system in which an alternating electric field is applied to hydrodynamically confined metallodielectric Janus particles, such as titanium-SiO_2. The latter are electrokinetically propelled along the electrode center line at a controllable velocity. When applied to large groups of particles, it is noted that co-moving ones move in single file, while those moving in opposite directions largely reorient and move past one another. At high particle densities, aggregates form as the many-body interactions become complicated. The particles can also make U-turns upon approaching the closed electrode end, but gradually slow at the open end and congregate in large piles. These behaviours can be explained largely in terms of a combination of an electro-osmotic flow into the electrodes and an asymmetrical electrical polarization of the Janus particles in the alternating electric field. Such observations can instruct strategies for manipulating micromotors: such as, for example, concentrating active particles at desired locations.

Self-propelling micromotors are a promising microscale tool for single-cell analysis and it has been shown[179] that the field gradients required to manipulate matter via dielectrophoresis can be induced at the surface of a self-propelling metallo-dielectric Janus particle under an externally applied electric field so that it essentially acted as a mobile floating micro-electrode. The use of the mobile floating micro-electrode to trap and transport cell organelles in a selective manner was extended here. The selectivity was ensured by differing dielectrophoretic potential wells on the particle surface and was controlled by the frequency of the electric field, together with hydrodynamic shearing and the size of the trapped organelle. This selective loading permitted the purification of relevant organelles from a mixed biological sample. Their dynamic release allowed their harvesting for the purposes of further analysis.

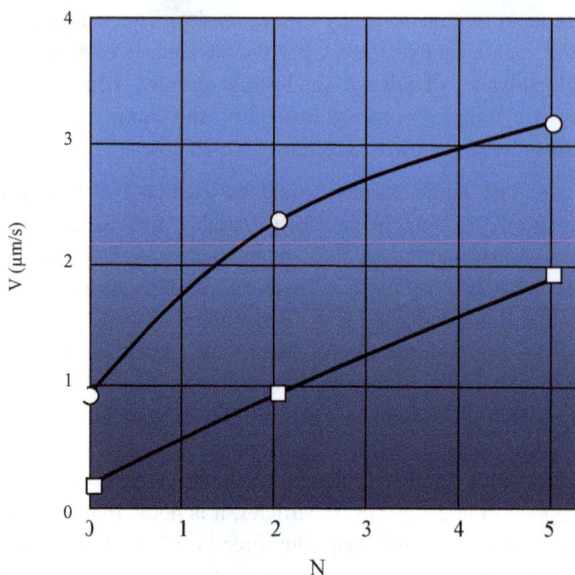

Figure 21. Velocity of disintegrating polymer particles as a function of the number of bilayers. Upper curve: silica particles coated with poly-dopamine followed by poly(L-lysine) and alternating layers of polyvinylpyrrolidone/poly(methacrylic acid). Lower curve: poly-dopamine-coated particles with poly(L-lysine) on one hemisphere and alternating deposition of polyvinylpyrrolidone/poly(methacrylic acid) on the other.

Other propulsive phenomena

The forces generated by electrochemical reaction can produce motion[180]. For example, when placed in solutions containing ions of the more noble metal, the spontaneous galvanic replacement of copper by platinum ions not only provides a sufficient electromotive force to propel the Janus particle but can also create asymmetrical platinum-sporting forms which may further function as catalytically propelled particles. Self-propelling particles may not even require the asymmetrical distribution of a catalyst, and simple isotropic particles can spontaneously form dimers which can self-propel. In a mixture of isotropic titanium dioxide particles which exhibit photochemical catalytic activity, and passive silica particles, illumination produces diffusiophoretic attraction between the active and passive components and thereby the formation of dimers. The latter are, in effect, broken-symmetry particles which can be controlled by the lighting conditions[181].

56

Materials Research Forum LLC
https://doi.org/10.21741/9781644901199

Although not strictly pertinent to archetypal Janus-particle motion, it is interesting to note that the pH-provoked disintegration of polymer multilayers has been found[182] to propel colloidal silica particles, with the velocities being higher for homogeneously-coated particles than for Janus particles. The particle velocity for both types of particle increases with increasing number of layers (figure 21), with increasing pH-gradient and with decreasing mass or diameter (figure 22). The existence of a steep pH-gradient leads to clearly directional motion whereas the motion of particles which are situated in a shallow pH-gradient is random. Rod-shaped and spherical particles which are of similar mass possess comparable mobilities. The observed directionality of the particles was attributed to chemotaxis, implying that this could constitute another means for controlling particles.

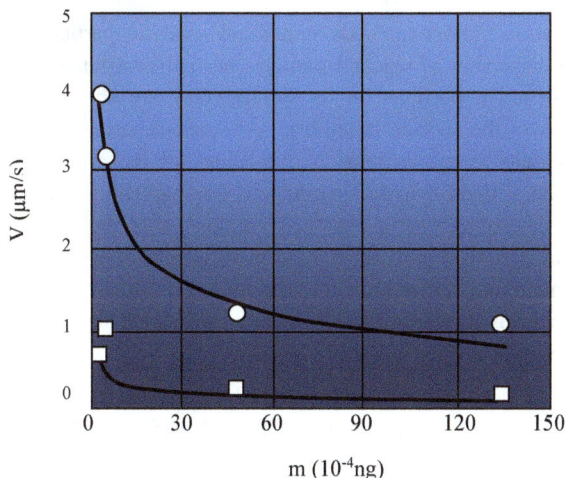

Figure 22. Velocity of disintegrating polymer particles as a function of particle mass. Upper curve: 5 layers of polyvinylpyrrolidone/poly(methacrylic acid). Lower curve: no layers

Metallodielectric Janus particles generally travel with the dielectric hemisphere facing forwards in a low-frequency applied electric field. This is attributed to an asymmetrical induced-charge electro-osmotic flow. At frequencies which are far above the charge relaxation time of the electric double layer which is created around the particle, the velocity does not decrease to zero: the particle instead reverses direction and moves with the metallic hemisphere facing forwards. This behaviour was attributed[183] to a surface force which arises due to localized non-uniform electric-field gradients which are in turn induced by other factors. One of these is the symmetry-breaking of an asymmetrical particle which is close to a wall that acts upon the induced dipole of the particle so as to produce motion, even when in a uniform alternating current field. Because the driving gradient is generated at the particle level, the latter is construed to be self-propelled even though the field is externally applied. This overall propulsion mechanism has been termed, self-dielectrophoresis; being distinct from conventional dielectrophoresis in which the external non-uniform driving field is governed from the outside and the particle direction is restricted. Both theory and experiment indicate that the critical frequency, at which the particle reverses direction is characterized by a non-dimensional parameter which incorporates the electrolyte concentration and the particle size.

Non-spherical osmotic motors are axisymmetrical catalytic particles which can self-propel in a dilute dispersion of reactant particles but unlike a Janus particle, which has reactive and non-reactive portions on its surface, they are able to move even when chemical activity over the surface is uniform. At small departures from sphericity, the self-propulsion velocity is proportional to the square of the distortion of the particle shape. The inclusion of hydrodynamic interactions can markedly affect self-propulsion in that, except in the case of very slow chemical reactions, even the direction of movement can change due to hydrodynamic interaction. The results of numerical calculations[184] suggest that the maximum self-propulsion velocity is exhibited by a sail-like shape. No plateau in the velocity is found, and the latter increases as the area of the area of the so-called sail increases and its thickness decreases.

Perhaps it should properly be included among the other externally imposed forms of propulsion, but it has been found[185] that a soft Λ-shaped non-buoyant particle can be made to move in a shaken liquid environment possessing zero mean velocity, and that this also applies to non-buoyant asymmetrical Janus microcapsules. Mathematical analysis indicates that the process requires symmetry-breaking and that this arises due to differing Stokesian drags acting during the 2 half-periods of oscillatory liquid motion. The natural anisotropy of the Λ-shaped particle or of the Janus capsule breaks the symmetry under sinusoidal liquid motion. The present mechanism again applies to more symmetrical soft particles if the symmetry is broken by imposing a suitable shaking action[186]. The swimming direction can also be selected by the choice of motion. Numerical studies have suggested that the propulsive mechanism can moreover counteract gravity, in that non-buoyant particles may be caused either to move upwards or to congregate at the container bottom.

Differential dynamic microscopic and particle-tracking studies[187] of the behavior of active Janus colloids reveal clear theoretical agreement with experimental data ranging from the scale where translational diffusion and self-propulsion predominate, up to the scale where effective diffusion is due to rotational Brownian motion. At intermediate length scales, oscillations bridge the changeover from directed motion to orientational relaxation and permit the discrimination of active Brownian motion and so-called run-and-tumble motion. Run-and-tumble motion[188] again mimics that of living entities and involves periods of rectilinear motion which are punctuated by sudden reorientations. The directional changes of synthetic active particles are due to a rotational diffusion which is superposed on a rectilinear motion and leads to relatively slow and continuous reorientations. But active particles can also exhibit motion wherein the translational and orientational changes are analogous to the run-and-tumble type. Such is the case for a viscoelastic solvent and a periodic modulation of the self-propulsion velocity, when light-activated Janus colloids are illuminated using a time-dependent laser. There is then a large increase in the effective translational and rotational motion when the modulation time is comparable to the relaxation time of the viscoelastic fluid. The behaviour is attributed to relaxation of the elastic stress which builds up during self-propulsion and which is suddenly released when the activity is turned off. Regardless of the propulsive mechanism of Janus particles, their guidance in the presence of thermal fluctuations remains difficult. Janus particles which were driven by using an electric field have been guided[189] by using a feedback-based system, and thus modelled the run-and-tumble behaviour of *Escherichia coli* but with deterministic steering being imposed during the tumbling phase. In a typical experiment, the Janus particle was set in the run state when its orientation vector was aligned with a target, and a transition to the steering state was triggered when the vector exceeded a specified tolerance angle. Deterministic reorientation of the particle was achieved by exploiting a characteristic rotational motion that could be switched on-and-off by modulating the alternating-current frequency of the electric field. The results showed that the feedback algorithm could be optimized by adjusting the tolerance angle. Other model run-and-tumble particles which exhibited combined translational and rotational self-propulsion were made up of doublets of Janus colloidal beads sporting catalytic patches that were positioned at a fixed angle with respect to one another[190]. The mean-square displacement and the mean-square angular displacement admitted of a simple Langevin description, from which characteristics such as the spontaneous translational and rotational velocities could be deduced. Two-dimensional feedback control like that above has been applied[191] to Janus particles which consisted of cobalt and platinum hemispheres self-propelling in hydrogen peroxide solution while directionally controlled by static magnetic fields. Because the magnetization direction of the particles could be heterogeneous, and inconsistent with the propulsion direction, they were useful for studying motional effects under 2-dimensional feedback control and open loop 3-dimensional control. Particles with their magnetization either closely aligned or very misaligned with respect to the propulsion direction could be caused to execute complex manoeuvres.

Although not strictly a new principle of propulsion, it is to be noted that the fuel which drives a particle need not be present in the medium *a priori*. The vapour-powered propulsion of catalytic particles is possible without making direct additions of fuel to the surrounding medium. The in-diffusion of hydrazine vapour from the atmosphere to the sample solution can provoke the rapid movement of iridium-gold Janus microspheres[192]. This mechanism was associated with a clear off-on response to the presence of fuel in the surrounding atmosphere, and with a spatio-temporal dependence of velocity which arose from the evolution of the concentration gradient within the solution.

Opto-thermoelectric microswimmers have been developed[193] which emulate the so-called run-and-tumble behaviour of *Escherichia coli* cells. The microswimmers were based upon dielectric-gold Janus particles, driven by an electrical field that arose from the asymmetrical optothermal response of the particles. Upon illumination with a defocused laser beam, the particles developed an optically-generated temperature gradient across the particle surfaces which led to an opto-thermoelectrical field which then propelled the particles. The swimming direction was governed by the particle orientation. Potential navigation of the swimmers was possible by affecting the in-plane rotation of the particles under a temperature gradient-induced electrical field produced by a focused laser beam. Timing the rotation laser beam would impart any desired orientation and thereby actively control the swimming direction. Dark-field optical imaging and feedback-control might achieve automated propulsion and navigation of the microswimmers. A promising navigation mechanism for fuel-based microswimmers such as autophoretic Janus particles involves modulating the local environment so as to guide the swimmer. One possibility is to etch grooves in microchannels, although such techniques have been limited to bulk guidance. It was been argued[194] that, by manufacturing microswimmers from phoretic filaments of flexible shape-memory polymer, elastic transformations can modulate the swimming behaviour, thus permitting precise navigation of selected individuals, within a group, through complex environments.

When the opposite boundaries of a narrow environment are maintained at a constant temperature, a non-uniform temperature profile can be set up across the film because of differential heat dissipation which is maximum at the boundaries[195]. Thermophobic particles then concentrate at the cooler boundaries while thermophilic particles concentrate in a layer which is midway between the boundaries. The Dufour-like effect leads to a synergism between the concentration and temperature profiles and amplifies the temperature gradient. A tenfold increase in particle concentration can occur rapidly with non-Janus particles at room temperature when using a low energy-input to maintain transverse temperature differences of just a few degrees Kelvin. Such a level of concentration was much higher than that predicted for Janus particles; the latter involving higher diffusion coefficients and a lower temperature-dependence. Unlike the case of Janus particles, such a system was expected to be stable and to exhibit both positive and negative thermophoresis.

One of the latest innovations has been to develop starch-based Janus particles[196]. A spin-coating method can produce both 12.2μm half-porous waxy cornstarch particles and 1.2μm half-hydrophobic amaranth starch particles. Untreated amaranth starch granules do not interact with each other in water, while more than 68% of the half-hydrophobic granules may self-assemble into worm-like strings and more than 66% of the fully hydrophobic granules aggregate into complex spherical supermicelles. The propulsive properties remain unexplored.

general analyses of Janus particle motion

Brownian aspects of the diffusiophoretic motion of spherical Janus particles were studied[197], showing that there existed 3 stages: simple Brownian motion at short times, superdiffusion at intermediate times and diffusive behavior at long times. These experimentally observed regimes were compared with a theoretical model for the Langevin dynamics of self-propelled particles exhibiting coupled translational and rotational motion. Theoretical predictions concerning the non-Gaussian behavior of self-propelled particles were verified. In agreement with Brownian dynamics simulations, an extremely broadened peak or a marked double-peak structure was found, depending upon the experimental conditions.

The Brownian transport of self-propelled overdamped particles in a 2-dimensional sectional channel was numerically investigated[198], showing that the resultant time-correlated Brownian motion was liable to rectification in the presence of spatial asymmetry. This ratcheting effect could be orders of magnitude greater for Janus particles; to such an extent that the autonomous pumping of a large mixture of passive particles could be achieved simply by adding a small fraction of Janus particles.

The hydrodynamic flow-field around a catalytically active colloid has been studied[199], by means of particle-tracking velocimetry, for both the free-moving state or when held stationary by an external force. The results yielded information concerning the fluid velocity in the vicinity of the colloid surface, and confirmed a proposed propulsion mechanism.

Entropic stochastic resonance was numerically simulated[200] for the case in which a self-propelled Janus particle moves in a double-cavity container. The results indicated that entropic stochastic resonance could persist even if there were no symmetry-breaking in any direction, thus demonstrating a key distinction between the behaviour of a self-propelled Janus particle and that of a passive Brownian particle, where symmetry-breaking is essential. The self-motion of reactive colloids and their dispersion behaviour have been examined theoretically[201]. This propulsion is not limited to Janus-like particles but can also occur with particles which have a uniform reactivity due to a more general mechanism: an entropic anisotropy created by the breaking of rotational symmetry. This situation was demonstrated with regard to the motion of a reactive particle due to a shape asymmetry or to the presence of an additional particle. In the two-particle case, sink or source particles could self-migrate towards or away from one another, respectively, at velocities which varied as R^{-2}, where R was the inter-particle distance. This thus resembled Coulomb attraction or repulsion. Because of this Coulomb-like behaviour, a suspension of sink particles could undergo collective flocculation, due to unscreened osmotic attraction. Such flocculation could occur when the particle volume fraction was within a certain range which in turn depended upon the solute concentration and the particle reactivity. The stability of reactive suspensions involved a modified Derjaguin-Landau-Verwey-Overbeek model which took account of the competition between long-range reaction-induced osmotic forces and short-range colloidal forces. The generalized view of self-driven particle motion was couched in terms of a simple scaling theory which offered a clear picture of the self-motion of two particles, composite bodies and Janus particles. All of the cases were driven by dipolar distortions of the potential energy.

By using a fluid particle-dynamics method a numerical investigation[202] was made of the motion which occurred due to the heating of half of a spherical Janus particle that was suspended in a binary liquid mixture. The method simultaneously accounted for the flow of solvent and for the motion of the particle, so that the velocity of the particle could be directly calculated. The analysis also took account of adsorption, in that a particle which was non-neutral with regard to adsorption was always entirely covered by an adsorption layer. In order to establish the self-propulsion mechanism, a 3-dimensional study was made of various combinations of adsorption preference for each hemisphere as a function of the heating power for symmetrical and non-symmetrical binary solvents and for various particle sizes. It was found that a reversal of the direction of motion occurred only for a particle in which the heated hemisphere was neutral while the other had a preference for one of the components of the binary mixture. The particle was predicted to self-propel much more rapidly in non-symmetrical binary solvents, and the self-propulsion could be traced to a gradient of mechanical stress which was distributed within the entire droplet rather than being localized. In this connection, a numerical investigation[203] was made of the propulsion of a Janus particle in a periodically phase-separating binary fluid mixture wherein the surface of the particle-tail preferred one of the binary fluid components while the particle-head was neutral with regard to wettability. During de-mixing, the more wettable phase was selectively adsorbed to the particle tail. Growth of the adsorbed domains led to hydrodynamic flow in the vicinity of the tail and this asymmetrical flow drove the particle in the direction of the particle head. During mixing, particle motion almost ceased because mixing occurred mainly via diffusion and the resultant hydrodynamic flow was then negligibly small. Repetition of the cycle continually moved the Janus particle towards the head.

Study of the temperature field and resultant flow pattern around a heated metal-capped Janus particle showed[204] that, if its thickness was greater than about 10nm, the cap was isothermal and the flow pattern incorporated a quadrupolar term which decayed as the inverse square of distance. In the case of much thinner caps, the velocity varied inversely as the cube of the distance.

The problems involved in the microscopic modelling of the chemical powering of Janus particles were discussed early on. Although continuum models could be used to describe the dynamics, models were required which could better account for intermolecular interactions, concentration gradients, thermal fluctuations and the generally complex out-of-equilibrium conditions. Analysis starting from first principles was needed. A microscopic model for the diffusiophoretic propulsion of Janus motors was therefore proposed[205] in which interaction with the environment occurred only through hard collisions. Phoretic mechanisms can be modelled[206] as involving the interconversion of 2 chemical species, and the situation has been analyzed in which the difference in chemical potential between the species can be taken to be constant, furnishing thermodynamically consistent equations of motion. Unlike the standard model of active Brownian particles having a constant self-propulsion velocity, this resulted in a non-constant velocity that depended upon the potential energy of the suspension. It was possible, by using this approach, to model consistently the breaking of detailed balancing and associated entropy production without invoking non-conservative forces.

A theoretical study[207] of the motion of a rigid dimer formed from self-propelling Janus particles used a simple kinetic approach with no hydrodynamic interactions. The dimer move in a helical trajectory while rotating about its center of mass. It was noted that inclusion of the effects of mutual advection did not change the qualitative aspects of the motion but altered only the trajectory and angular velocity parameters.

An efficient method has been developed[208] for the direct simulation of the motion of non-equilibrium particles without requiring excessive computation of the complex multi-timescale problem. Fluctuating lattice Boltzmann methods and grid-refinement techniques were combined in order to simulate spherical and cylindrical Janus particles. Comparison of their motional characteristics with experimental data at multiple time-scales proved that the method was valid.

A theoretical study which could encompass both diffusio- and electro-phoretic phenomena has been made[209] of the self-diffusiophoretic motion of a Janus particle by using the Onsager-Casimir reciprocal relationships. Rectilinear and angular velocities of a single Janus particle were shown to result from coupling of electrochemical forces to the fluid flow-fields induced by a force and to torque on the particle, respectively. A typical example considered a half-capped particle, which catalyzed the chemical reaction of solutes at its surface, by reducing the continuity equations of the reacting solutes to Poisson equations which described the corresponding electrochemical fields. Anisotropic surface reactivity alone was shown to be sufficient to produce linear motion of a Janus particle. Rotation occurred only if the particle was not axisymmetrical. In the absence of interaction with the solute, the linear velocity of the particle was related to a randomly weighted sum of the frictional coefficients or of the hydrodynamic radii of the reactive solutes. Interaction with the solutes was certainly required in order to result in far-field diffusiophoretic interactions between Janus particles if they involved an interfacial solute excess at the particle surface.

The motion of a Janus particle self-propelling phoretically in a weakly viscoelastic fluid has been modelled[210] numerically. The self-propulsion generally involves an asymmetry, in the properties of the surface of a Janus particle, which then drives a surface slip velocity and bulk flow. The effect of viscoelasticity on this advection-diffusion phenomenon was investigated for a range of Péclet and Damköhler numbers. It was found that particles could move faster, or slower, in viscoelastic fluids. Reaction and diffusion rates affected the viscoelastic stresses which led to changes in propulsion. An experimental investigation has been made[211] of the motion of spherical Janus particles in a viscoelastic fluid, with self-propulsion being provided by a local concentration-gradient of a polymer mixture, which was imposed by laser illumination. Even when the fluid viscosity was independent of the deformation-rate imposed by the particle, there was an increase of up to 2 orders of magnitude in rotational diffusion with increasing particle velocity. The data could be described in terms of an effective rotational diffusion coefficient which depended upon the Weissenberg number. The overall effect was to produce a very anisotropic response, to external forces, depending upon the orientation of the particle.

A large-scale molecular dynamics simulation was made[212] of the orientational dynamics of a heated self-propelling Janus particle. The asymmetry in the microscopic interaction of the particle with the solvent was incorporated by imposing differing wetting parameters for the two hemispheres. This led to a differing microscopic Kapitza resistance across the solid/fluid boundary of the 2 hemispheres and thus a gradient in temperature was created across the poles. This temperature gradient led to self-propulsion along the direction of the symmetry axis. The orientational correlation of the symmetry axis was deduced from the simulation and led to a measure for the rotational diffusion constant: heating led to an increase in the rotational diffusivity. The increase in rotational diffusion was related to the temperature difference across the poles of the sphere and to the average surface temperature difference with respect to the ambient fluid. Because the rotational diffusion was governed by the overall flow-field in the ambient, it was deemed that comparison of the rotational diffusion with the temperature difference across the poles was misleading and that comparison with the temperature difference with respect to the ambient was better. The latter choice resulted in the ability to collapse the data for various microscopic interactions. The directionality of the self-propulsion was shown to change, depending upon the microscopic interaction. When attractive interaction of the colloid with the solvent was switched off, the phoretic mobility changed sign. The propulsion velocity was zero if the heating was below a threshold value.

The particular case represented by Janus colloidal particles situated at the water surface has been analysed[213]. Autonomous motion in general occurs when they can react chemically with the liquid in which they are immersed. In order to understand the self-propulsion of catalytic Janus colloids at the air/water interface, both wetting aspects and the orientation of the catalytic surface have to be considered. Wetting governs contact between the fuel and the catalytic surface, and motion is not likely to occur if the catalytic face is entirely out of the water or when the Janus boundary is parallel to the surface. A Janus colloid having 2 hydrophilic faces must permit the catalytic face, such as platinum, to react with, for example, H_2O_2 in the water. It must also permit rotational freedom in order to generate propulsion parallel to the surface. The free energy of an ideal Janus particle at the interface has been determined as a function of immersion depth and particle orientation. By introducing the possibility of contact-angle hysteresis, it is possible to describe how the effects of contact-line pinning modify the ideal case. In fact, experimental observations of the contact-angle hysteresis of Janus colloids at the interface reveal that the orientations of silica particles which are half-covered with platinum, when at the surface, do not agree with the ideal case.

Given that the mobility of Janus particles is their primary advantage, the 1-dimensional and 2-dimensional diffusion-limited reactions, $A + A \rightarrow 0$ and $A + B \rightarrow 0$, have been considered[214] where A was an active Janus particle and B was a passive particle; all being in thermal equilibrium. Upon increasing the self-propulsion time of the A-particles, the reactant densities decayed faster within the regime of time-transients which was of likely relevance to chemical applications. The asymptotic and transient density-decreases obeyed power-laws, the exponents of which depended upon the nature of the reaction and its dimensionality.

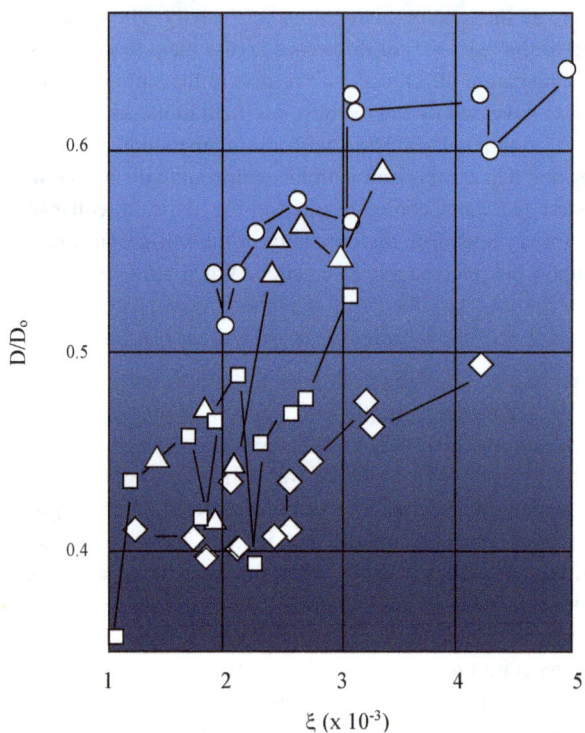

Figure 23. Diffusivity of active chains, normalized with respect to the diffusivity of a single bead, as a function of flexibility. Circles: 3 particles in chain, diamonds: 4 particles in chain, squares: 5 particles in chain, diamonds: 6 particles in chain

Colloidal suspensions can self-assemble into dense clusters when doped with Janus particles. The clusters may temporarily move in the direction opposite to that of the self-propulsion direction of the resident Janus particles. A model has been developed[215] which accounts for such a velocity-reversal as well as for other aspects of cluster behaviour. The reversal was attributed to a non-reciprocal phoretic attraction of the passive particles, to the caps of the active particles, which occurred even when in close contact and which pushed the active particles backwards. When the phoretic interactions were repulsive they led to a dynamic aggregation of passive colloids at the chemical density minima produced by the active particles. Under some conditions, this could lead to the appearance of moving fronts of active particles which were pursued by passive ones.

As another example of the manner in which synthetic particles mimic living creatures, it can be argued in principle that one Janus particle can even communicate with another and affect its propulsion behaviour[216]. A chemical message can be sent from a moving particle, to another nearby particle, by the release of Ag^+ ions from a polystyrene-Ni/Au/Ag Janus particle to a platinum-SiO_2 particle. The cloud of silver ions causes a marked velocity change which is associated with an increased catalytic activity. Selective activation of multiple particles is possible via sequential localized chemical communications. At high concentrations, in fact, interactions between particles can lead to complex emergent behaviours, in which collective dynamics result in the appearance of groups comprising hundreds of members. There can occur a marked increase in diffusivity for particles which are of Janus type or which are present in high concentrations. Uniformly catalyst-coated particles were linked together[217] in a chain in order to guarantee their closeness. Changes arose due to hydrodynamic interactions between the linked particles, due to reaction-induced phoretic flows that were catalyzed by platinum on their surface. This resulted in diffusivity increases of up to 60% for individual chains. Those chains possessing greater flexibility exhibited higher diffusivities (figure 23). Simulations suggested that the diffusivity increase was due to an interplay between conformational fluctuations of the chain and activity.

In general terms the self-propulsion of synthetic motors, like camphor-boats and phoretic Janus particles, involves the production of various chemicals; but the same chemicals which provoke the self-propulsion of a given Janus particle can also act upon others and move them towards or away from the first particle[218]. Single mobile particles leave chemical trails behind them with which they interact, leading to self-trapping or self-avoidance. When they are part of an ensemble of mobile particles, each one also responds to chemicals produced by the others. Attractive interactions usually lead to clustering, even at low particle densities. Those clusters can either progress towards macrophase separation, or towards dynamic clusters of self-limited size as in the case of autophoretic

Materials Research Forum LLC
https://doi.org/10.21741/9781644901199

Janus particles. Rarer chemical interactions can create novel patterns in active systems and lead to patterns, such as clusters, which are surrounded by shells of chemicals, travelling waves and more complex continually re-forming patterns. Swarms of glucose-oxidase-powered gold Janus particles exhibited positive macroscale chemotaxis[219]. The gold nanoparticles were rendered asymmetrical by grafting polymer brushes onto one side, with glucose oxidase being present on the other side. The resultant particles exhibited velocities of up to some 120 body-lengths/s in the presence of glucose. The polymer brushes greatly increased the translational diffusion of the Janus gold particles, and they also exhibited collective chemotactic progress along the concentration gradient of a glucose source.

Because their translation depends upon the fuel concentration, it is expected that active colloidal particles should be able to swim towards a fuel source. The engineering of nanoparticles having distinct chemotactic properties might permit the development of particles which can autonomously swim along a pH gradient towards a tumour. Chemotaxis requires that particles should possess an active coupling of their orientation to a chemical gradient. A simple description was provided[220] for the underlying mechanisms of chemotaxis, as well as means for analyzing active particles that can exhibit positive or negative chemotaxis.

It was shown that chemical molecules dissolved in aqueous suspensions can mediate long-range interactions and thus stabilize a polar phase. Chemo-attractants in living suspensions, and dissolved molecules in synthesized Janus suspensions, resemble such chemical molecules. Communication between swimmers via their chemical gradients is the basis of this stabilization mechanism. Detailed phase diagrams were provided[221] for two classes of systems involving momentum-conserving and non-conserving dynamics. A linear stability analysis showed that the proposed stabilization mechanism could work for swimmers, both pushers and pullers, having spherical, oblate or prolate geometries.

A molecular-dynamics model of a Janus particle driven by chemotaxis had already been developed[222]. Upon increasing the intensity of the stimulus, it was predicted that transport of the particle exhibited a second-order state transition from a composite random walk to increased directional transport, with a size-dependent reversal of motion. The prior random walk also comprised power-law distributed truncated Lévy flights and Brownian jiggling. A state diagram of Janus-particle transport neatly summarised the combined effect of stimulus intensity and particle size (figure 24). The physical mechanisms involved in the transport behaviours were explored by theoretical modelling which was based upon noise and the Janus geometry.

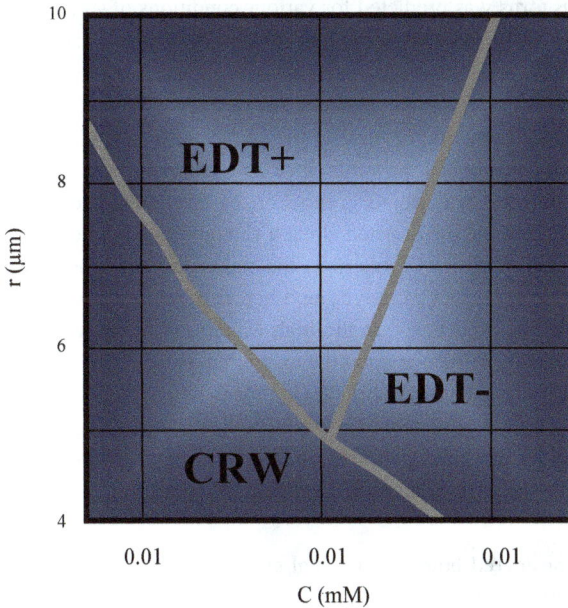

Figure 24. Concentration-diameter state diagram for the predicted chemotactic behaviour of Janus particles. EDT+: enhanced directional transport towards higher concentrations, CRW: composite random walk, EDT-: enhanced directional transport towards lower concentrations.

Chemically active toroidal colloids behave differently to spherical colloids, and coarse-grained microscopic simulations of the dynamics of self-diffusiophoretic toroidal colloids have been used[223] to study how geometrical factors affect their motion. The concentration and velocity fields around the colloid are sensitive functions of parameters such as the torus hole size and the thickness of the torus tube. Toroidal colloids having various geometrical and dynamic characteristics could be created, and studied in fluids which involved chemical reaction and fluid flows. Earlier work[224] had considered the flow which was induced by an axisymmetrical phoretic torus. The solution was in the form of an infinite series solution which was confirmed using boundary-element computations. For a torus of uniform chemical activity, confinement effects in the hole caused the torus to act as a pump. This was optimized, subject to a fixed particle surface area. The fastest-swimming Janus torus was predicted for various conditions of surface chemistry. None of the optimal tori clearly occurred in the limit of zero central hole size.

gravitaxis

The asymmetrical distribution of mass at the surface of a catalytic Janus particle tends to result in its being propelled upwards in a gravitational field. The existence of this gravitaxis phenomenon was noted[225] early on by observing the trajectories of fuelled Janus particles, where the thrust acted along a vector which pointed away from the metal-coated hemisphere. With increasing size, the trajectories of spherical particles were no longer isotropic with respect to gravity and began to move very preferentially upwards. This was attributed to the fact that the high-density asymmetrical platinum caps had a size-related increasing influence upon the azimuthal angle of the Janus sphere and turned its orientation towards a configuration having the heavier, propulsive, surface face down. There was in fact good agreement between the distribution of experimentally observed azimuthal angles for the Janus particles, and the predictions offered by simple Boltzmann statistics. Gravitaxis is thus another potential mechanism for the control of catalytic particles.

Later work[226] considered both translational and rotational motion, again showing that in the case of particles having an anisotropic mass distribution motion is generally downward. Large mass anisotropies lead to particle motion along helical trajectories, in which the axis is oriented parallel, or antiparallel, to the gravitational force. When the mass anisotropy is small, and rotational diffusion predominates, no gravitational alignment of the trajectory occurs. The observed trajectories depend upon the angular self-propulsion velocity of the particle. If this component is large, and rotates the direction of translational self-propulsion of the particle, the trajectory has many loops. Only elongated periods of travel occur if the angular self-propulsion is low. The gravitational alignment mechanism, and the dependence of trajectory shape upon the angular self-propulsion, could be used to separate active particles with regard to mass anisotropy and angular self-propulsive ability, respectively.

effect of barriers

The interaction of the particles with various obstacles is perhaps one of the most intriguing and important aspects of their behaviour, as such interactions will determine their practical usefulness. While various mechanisms can produce self-propulsion, the ability to steer self-propelled entities has been more problematic. Their motion has to counter the Brownian rotation which randomizes the particle orientation. The directed motion of catalytic motors moving close to solid surfaces is useful here[227] and is achieved by the active quenching of their Brownian rotation by constraining it in a rotational well arising from hydrodynamic effects. The combining of these geometrical constraints can steer active colloids along arbitrary trajectories. Some of their possible extensive applications[228] will require that the particles sense and react to the local environment in a reliable manner. The self-generated hydrodynamic and chemical fields, which induce particle motion, can detect and be modified by the environment; including confining boundaries. Considering a catalytically active Janus particle to be a typical example, it was predicted that - near to a hard planar wall - such a particle could exhibit various forms of motion: reflection from the wall, motion with a steady-state orientation and height above the wall, or a motionless steady so-called hovering behaviour. In these steady states, the height and orientation are governed by the proportion of catalyst coverage and the interaction of the solutes with the differing faces of the particle. It is assumed that a desired behaviour can be obtained by adjusting the parameters via control of the particle surface chemistry.

A density functional theory investigation was made[229] of the structure formation of amphiphilic molecules at planar walls. The molecules were modelled as hard spheres which were composed of a hydrophilic and a hydrophobic part. The orientation of the resultant Janus particles was described by a vector representing an internal degree of freedom. A density functional approach involved fundamental measure theory combined with a mean-field approximation for the anisotropic interaction. Assuming an environment of neutral, hydrophilic and hydrophobic walls, a study was made of particle adsorption with particular attention being paid to competition between the surface field and interaction-induced ordering. In systems which are confined between two planar walls, an anisotropic Janus interaction can produce marked frustration effects at low temperatures. The trajectories of self-propelled organisms or synthetic devices in a viscous fluid can thus be altered by hydrodynamic interactions with nearby boundaries.

Brownian dynamics simulations have been used[230] to predict the translational and rotational fluctuations of a Janus sphere, near to a boundary, having a cap of non-matching density. That is, the presence of the cap markedly affected the rotational dynamics of the particle, due to a gravitational torque, under experimentally relevant conditions. Gravitational torque predominated over stochastic torque for particles which were greater than 1μm in diameter and had a 20nm-thick gold cap. Janus particles under these conditions had largely cap-down orientations. Although the results showed that particles having diameters of less than 1μm and a gold coating thinner than 5nm behaved like an isotropic particle, small increases in the particle diameter or coating thickness quenched the polar rotation of the particle. Particles having larger diameters or thicker gold coatings had mainly cap-down configurations.

Materials Research Forum LLC
https://doi.org/10.21741/9781644901199

Three-dimensional/two-component microparticle image velocimetry was used[231] to examine the hydrodynamic flow patterns around 15μm-diameter metallodielectric Janus particles adjacent to insulating and conducting walls. When far from the walls, the observed flow patterns were in good qualitative agreement with experiment. When close to a conducting wall, strong electrohydrodynamic flows occurred at low frequencies, resulting in fluid being injected towards the particle. Closeness of a metallic hemisphere to a conducting wall also produced a localized field gradient, resulting in the dielectrophoretic trapping of 300nm polystyrene particles over a wide frequency-range.

A multipole description of swimming bodies can provide[232] a general framework for the study of fluid-mediated modifications to particle trajectories. A general axisymmetrical swimmer can be described as being a linear combination of fundamental solutions to the Stokes equations: that is, a Stokeslet dipole, a source dipole, a Stokeslet quadrupole and a rotlet dipole. The trajectory-affecting influence of nearby walls or stress-free surfaces can be described with respect to the contribution of each singularity and is expected to reflect the wall effects observed in complete numerical solutions of the Stokes equations. A reduced model could furnish simple and accurate predictions of wall-induced attraction and dynamics of model Janus particles, cilia-equipped micro-organisms and polar bacteria-like entities. In one investigation, a density functional theory study was made[233] of the structure and adsorption of amphiphilic molecules at planar walls which were modified by tethered chain molecules. The molecules were modelled as spheres comprising a hydrophilic part and a hydrophobic part. The pinned chains were treated as tangentially-joined spheres which could interact with the fluid molecules via orientation-dependent forces. This density functional approach took advantage of fundamental measure theory, thermodynamic perturbation theory for chains and a mean-field approximate description of anisotropic interactions. In studying particle adsorption, emphasis was placed on the competition between external fields, due to the surface and attached chain molecules, and interaction-induced ordering.

Numerical simulations have been made[234] of the transport of elliptical Janus particles along narrow two-dimensional channels having reflective walls. The self-propulsion direction of the particle could be aligned along either its major (prolate) or minor (oblate) axis. In the case of smooth channels, there were long diffusion transients. These were ballistic in the case of prolate particles and exhibited zero diffusion in the case of oblate particles. When situated in a rough channel, the prolate particles tended to drift against an applied driving force by tumbling over wall protrusions. For some aspect ratios, the degree of negative mobility could become very high and was then termed giant negative mobility. It was expected that a modest external driving force was sufficient to ensure the self-propulsion of rod-like Janus particles in rough channels.

A further numerical investigation was made[235] of the escape kinetics of elliptical Janus particles from narrow two-dimensional cavities having reflecting walls. The self-propulsion velocity of the Janus particle was again directed along either its major (prolate) or minor (oblate) axis. The mean exit time was very sensitive to the cavity geometry, particle shape, and self-propulsion strength. The mean exit time was at a minimum when the self-propulsion length was equal to the cavity size. The optimum mean escape time could be found as a function of the self-propulsion velocity, translational diffusion and particle shape and this could aid the control of Janus particles in a channel.

In the case[236] of Janus particles diffusing along narrow periodically -corrugated channels, The swimmer self-propulsion mechanism could be modelled so as to incorporate a non-zero torque (propulsion chirality)[237]. With regard to chirality, left-right and inverted asymmetrical channels could exhibit differing transport properties. The self-propulsion torque was found to play a pivotal role in transport control. Further numerical simulations[238] were made of the diffusion of overdamped point-like Janus particles along narrow two-dimensional periodically corrugated channels having reflecting walls. The self-propulsion velocity of the particle was assumed to rotate due to an imposed torque. When the mirror symmetry of the channel was broken with respect to its axis, this was sufficient to generate a directed particle flow which exhibited an orientation and velocity which depended upon the channel geometry and the particle-locomotion properties. Chiral microswimmers thus drifted autonomously along a narrow channel under asymmetry conditions which were more general than previously suspected. An investigation was made[239] of the transport diffusivity of Janus particles, in the absence of external biases. The case was considered of chiral Janus particles which moved either in the bulk, or in sinusoidal channels having reflecting walls. Their self-diffusion constants depended upon both the strength and chirality of the self-propulsion mechanism. In a periodic channel, self-diffusion could be controlled by tailoring the compartment geometry.

There also exist chiral active particles, which are also known as self-propelled circle swimmers. These include sperm cells and asymmetrical Janus colloids, and exhibit a wealth of patterns which differ from those formed by linear swimmers. These patterns have usually been explored for identical circle swimmers, while real-world circle swimmers typically have a frequency distribution. It was shown[240] that even the simplest mixture of velocity-aligning circle swimmers, having two different frequencies, yielded a complex set of superstructures. The most remarkable of these involved a micro-flock pattern of one species, while the other species phase-separates to form a macrocluster which coexisted with a gas phase. Here, one species microphase-separated at a characteristic length scale while the other macrophase-separated to give a chosen density. Another interesting case occurred in an isotropic system, yielding patterns which comprised two different characteristic length-scales that were controllable via the frequency and swimming speed of the individual particles.

An investigation was made[241] of the active motion of self-propelled colloids confined at the air-water interface and explored the possibility of enhancing the directional motion of self-propelled Janus colloids by slowing down their rotational diffusion. The two - dimensional motion of micron-sized silica-platinum Janus colloids has been experimentally measured by particle -tracking video-microscopy at increasing concentrations of the catalytic fuel: that is, hydrogen peroxide. When compared to the motion occurring in the bulk, a dramatic enhancement of both the persistence length of trajectories and the speed is observed. The interplay of colloid self-propulsion, due to an asymmetrical catalytic reaction occurring on the colloid, surface properties and interfacial frictions controls the enhancement of the directional movement. The slowing down of the rotational diffusion at the interface, also measured experimentally, plays a pivotal role in the control and enhancement of active motion.

An investigation has also been made[242] of the motion of silica-palladium particles which were adsorbed at the interfaces between nematic liquid crystals and aqueous phases which contained hydrogen peroxide. The elasticity and anisotropic viscosity of the nematic phase changed the particle motion at the liquid crystal interfaces. Contact line pinning on the particle surface restricted their out-of-plane rotational diffusion at the interfaces, while orientational anchoring of nematic liquid crystals on the planar silica and homeotropic palladium hemispheres biased the particle in-plane orientations so as to produce almost motion which was almost exclusively aligned along the director of the liquid crystal at 0.5wt% peroxide concentrations. The displacements perpendicular to the director were those of Brownian diffusion. At 1 to 3wt% peroxide, an increasing fraction of particles moved parallel to and perpendicular to the liquid crystal director. Under these conditions, some 10% of the particles moved only perpendicular to the liquid crystal director.

A study was made[243] of the effect of a nearby planar wall on the propulsion of a phoretic Janus micro-swimmer driven by asymmetrical reactions, on its surface, which absorbed reactants and generated products. The behaviour of these swimmers near to a wall could be distinguished on the basis of whether the swimmers were mainly absorbing, or producing, reaction solutes and whether their swimming directions were such that the inert or active portion faced forwards. The wall-induced solute gradients then always promoted swimmer propulsion along the wall, while hydrodynamic effects led to re-orientation of the swimming direction away from the wall.

In the case of the self-propulsion of 2μm -diameter platinum-SiO$_2$ Janus microspheres near to a wall, measurements were made[244] of the relationship between the self-propellant velocity, V$_{Janus}$, as a function of the observed time, Δt$_{obs}$. A diffusiophoretic force-dominated motion, which was quasi one-dimensional and which was both force -free and torque -free, could be distinguished from the overall motion. The Janus microsphere was also deflected about the vertical direction by a deflection-angle, ψ. This decreased with increasing H$_2$O$_2$ concentration. For solutions with 2.5 to 10% of H$_2$O$_2$, ψ ranged from 20° to 7°. A numerical model which took account of the viscous force, diffusiophoretic force and effective gravity, could satisfy the conditions of force and torque balance simultaneously for the case of quasi one-dimensional self-propellant motion. Changes in ψ, and in the separation distance, δ, of microspheres from the substrate were studied for various values of H$_2$O$_2$ concentration, material density and microsphere diameter. In the case of the self-propulsion velocity, V$_{Janus}$, and the deflection angle, ψ, the numerical results agreed well with experimental observations. A lower density or a smaller microsphere diameter led to a smaller δ, while a higher H$_2$O$_2$ concentration led to a larger δ. The predicted value of the latter was 2 to 8μm. The near-wall effect upon the characteristic time, τ$_R$, of rotational diffusion of the Janus microsphere could then be examined. Because of the relatively high predicted values of δ, the near -wall effect was not expected to cause a large discrepancy in τ$_R$.

The dynamics of self-propelled Janus particles interacting with catalytically passive silica beads in a narrow channel have been studied[245] via experiment and numerical simulation. Upon varying the areal density of the beads and the channel width, the resultant changes ranged from distinct bulk and boundary-free diffusivities at low densities, to directional so-called locking and channel unclogging at higher densities. This confirmed that a Janus particle was capable of shepherding large clusters of passive particles. Computer simulation of the behaviour of a mixture of large passive charged colloids within a suspension of smaller active dipolar Janus particles showed[246] that, when a single charged colloid was present in solution, it acquired a rotational or translational motion which depended upon how the active dipoles self-assembled on its surface to form active complexes. The collective behavior of the complexes included swarming, and coherent macroscopic motion. These behaviours could be controlled by varying the strength of the active forces and the relative concentrations of the species. A theoretical study had been made[247] of the motion of surface-active Janus particles which were driven by an effective slip velocity that was due to a non-uniform temperature or concentration field. Using parameters which were relevant to thermal traps, it was found that the torque which was exerted by the gradient of the field inhibited rotational diffusion and favored alignment of the particle axes. In the case of a swarm of active particles, such polarization could add a new term to the drift velocity and modify the collective behavior. Self-polarization in a non-uniform laser beam could be used to guide hot particles along a chosen trajectory.

In general, Janus phoretic colloids self-propel due to self-generated chemical gradients and exhibit spontaneous non-trivial dynamics, in suspensions, at length-scales which are much greater than their own size. Collective dynamics then arise from a competition between the self-propulsion velocity of the particles, the chemically-mediated interactions between particles and the disturbances introduced into the flow of the surrounding medium. The factors are directly governed by the shape and physicochemical properties of the colloid surface. Using a popular kinetic model for dilute suspensions of a chemically active species having far-field hydrodynamic and chemical fields, a linear stability analysis showed[248] that self-propulsion could involve a wave-selective mechanism for certain particle configurations. Numerical simulations showed that the appearance of regular patterns in the particle density was promoted by chemical interactions but was prevented by the high fluid flows that were collectively generated by polarized particles, whether or not they were puller or pusher swimmers.

A study was made[249] of the fission and fusion of magnetic microswimmer clusters, as governed by hydrodynamic and dipolar interactions. The results depended upon whether the swimmer was a pusher or a puller: a linear magnetic chain of pullers was stable whereas a pusher chain underwent a cascade of fission processes as the self-propulsion velocity was increased. Magnetic ring clusters exhibited fission for either type of swimmer. Numerous possible fusion or assembly events were possible if a single swimmer collided with a ring-like cluster or two rings spontaneously collided. The predictions were verifiable by experiments performed on active colloidal Janus particles and magnetotactic bacteria.

Simulations of the trajectory of a Janus sphere having a cap density which matched that of the base particle very near to a boundary were used to construct 3-dimensional potential -energy landscapes that could then be used to deduce particle and solution properties. The Results showed that the potential energy landscape of a Janus sphere exhibited a transition region, at the boundary between the two Janus-related halves, which depended upon the relative zeta -potential magnitude. The potential energy landscape was fitted so as to deduce accurately the zeta potential of each hemisphere, the particle size, the minimum potential energy position and electrolyte concentration, and the Debye length. It was possible to determine the appropriate orientation bin size and regimes over which the potential energy landscape had to be fitted in order to obtain the system properties. The study unfortunately revealed that an experiment could well require more than a million observations to be performed in order to obtain a suitable potential energy landscape. This excessive number was dictated by the multivariable observations which were necessitated by an anisotropic particle.

An experimental and theoretical study[250] was made of the dynamics of chemically self-propelled Janus colloids moving atop a two-dimensional crystalline surface. The surface was an hexagonally close-packed monolayer of colloidal particles having the same size as that of the mobile one. The dynamics of the self-propelled colloid was driven by a competition between a hindered diffusion due to the periodic surface, and an increased diffusion due to active motion. The predominant contribution depended upon the propulsion strength, and this could be adjusted by changing the fuel concentration. The measured mean-square displacement revealed the occurrence of increased diffusion at long lag -times. The experimental data could be described by a Langevin model for two-dimensional translational motion of a Brownian particle in a periodic potential that combined the effects of gravity and the crystalline surface, but allowed free rotational diffusion of the colloid. There was a crossover from free Brownian motion at short times to active diffusion at long times.

Chemically active colloids move by creating gradients in the composition of the surrounding solution and by exploiting the differences in their interactions with the various molecular species in solution. If such particles move near to boundaries, e.g. the walls of the container confining the suspension, gradients in the composition of the solution are also created along the wall. This gives rise to chemi-osmosis (via the interactions of the wall with the molecular species forming the solution), which drives flows coupling back to the colloid and thus influences its motility. Employing an approximate 'point-particle' analysis, it was shown[251] analytically that - owing to this kind of induced active response (chemi-osmosis) of the wall - such chemically active colloids can align with, and follow, gradients in the surface chemistry of the wall. In this sense, these artificial 'swimmers' exhibit a primitive form of thigmotaxis with the meaning of sensing the proximity of a (not necessarily discontinuous) physical change in the environment. Alignment with the surface-chemistry gradient is generic for chemically active colloids as long as they exhibit motility in an unbounded fluid. That is, this phenomenon does not depend upon the exact details of the propulsion mechanism. The results were explained within the context of simple models of chemical activity, corresponding to Janus particles with 'source' chemical reactions on one half of the surface and either 'inert' or 'sink' reactions over the other half.

The achievement of control over naturally diffusive out-of-equilibrium systems composed of self-propelled particles, such as cells or self-phoretic colloids, is a long-standing challenge in active -matter physics. The inherently random motion of these active particles can be rectified in the presence of local and periodic asymmetric cues, given that a non-trivial interaction exists between the self-propelled particle and the cues. The phoretic and hydrodynamic interactions of synthetic micromotors with local topographical features have been exploited[252] in order to break the time-reversal symmetry of particle trajectories and to direct a macroscopic flow of micromotors. The orientational alignment induced on the micromotors by the topographical features, together with their geometrical asymmetry, is crucial in generating directional particle flow. The system can be used to concentrate micromotors in confined spaces and to identify the interactions leading to this effect. A minimal model identified the key parameters of the system responsible for the observed rectification. The system allows for robust control over both the temporal and spatial distribution of synthetic micromotors.

Self-phoretic chemically active Janus particles move by inducing - via non-equilibrium chemical reactions occurring on their surfaces - changes in the chemical composition of the solution in which they are immersed. This process leads to gradients in chemical composition along the surface of the particle, as well as along any nearby boundaries, including solid walls. Chemical gradients along a wall can give rise to chemi-osmosis, i.e., the gradients drive surface flows which, in turn, drive flow in the volume of the solution. This bulk flow couples back to the particle, and thus contributes to its self-motility. Since chemi-osmosis strongly depends upon the molecular interactions between the diffusing molecular species and the wall, the response flow induced and experienced by a particle encodes information about any chemical patterning of the wall. Previous studies of the self-phoresis of a sphere near to a chemically patterned wall were extended to the case of particles having a rod-like elongated shape. Analysis was focused[253] upon new phenomenology which could potentially emerge from the coupling - which was inoperative for a spherical shape - of the elongated particle to the strain rate tensor of the chemi-osmotic flow. Detailed numerical calculations showed that the dynamics of a rod-like particle exhibited a novel 'edge-following' steady -state: the particle translates along the edge of a chemical step at a steady distance from the step and with a fixed orientation. Within a certain range of system parameters, the edge-following state co-exists moreover with a 'docking' state (the particle stops at the step, oriented perpendicular to the step edge), i.e., a bistable dynamics occurs. These findings are rationalized as being a consequence of the competition between the fluid vorticity and the rate of strain by using analytical theory based upon the point-particle approximation which quasi-quantitatively captures the dynamics of the system.

For natural microswimmers, the interplay of swimming activity and external flow can promote robust directed motion, for example, propulsion against (upstream rheotaxis) or perpendicular to the direction of flow. These effects are generally attributed to their complex body shapes and flagellar beat patterns. Using catalytic Janus particles as a model experimental system, a strong directional response that occurs for spherical active particles in a channel flow was reported[254]. The particles align their propulsion axes so as to be nearly perpendicular to both the direction of flow and the normal vector of a nearby bounding surface. A deterministic theoretical model of spherical microswimmers near to a planar wall was developed which described the experimental observations. The directional response arose from the interplay of shear flow and near-surface swimming activity. Upon adding the effect of thermal noise, probability distributions were deduced, for the swimmer orientation, which tended to agree with the experimental distributions.

Experiments have been performed on SiO_2-Pt Janus particles which were suspended in an aqueous medium within a capillary that was subjected to various shear flow rates[255]. The particles were propelled by various H_2O_2 concentrations. At a given speed, a continuous transition in the motion of the particles occurred with increasing shear flow: from the usual random motion, to preferential motion along the vorticity direction and thence to migration along the flow. The transition was associated with a marked decrease in the in-plane fluctuations of the particle trajectories. The activity of the particles produced a delay in shear-induced rolling. At moderate flow-rates, this permitted the particles to adopt a specific orientation and aided their migration along the vorticity direction. At higher flow rates, once the shear flow had overcome the activity-induced resistance and initiated rolling, the likelihood that the particles could take up a preferred orientation decreased. The transitions were suggested to be determined by a non-dimensional parameter which involved the relative strength of the shear-induced particle flow and the propulsion speed.

If catalytically active Janus particles are dispersed in certain liquid solutions, they can create a gradient in the chemical composition of this solution along their surfaces, as well as along any nearby confining surfaces. This gradient drives self-propulsion via a self-phoretic mechanism, while the compositional gradient along a wall gives rise to chemiosmosis, which additionally contributes to self-motility. In one study, a theoretical analysis was made of the dynamics of an active colloid near to chemically patterned walls. A point-particle approximation combined with a multipole expansion was used[256] to explore the effects of pattern geometry and chemical contrast upon the particle trajectories. Particular attention was paid to planar walls which were patterned with chemical steps and stripes. Changes in the topology of the corresponding phase portraits occurred upon varying the chemical contrast and the stripe -width.

Active colloidal particles regularly interact with surfaces in applications ranging from microfluidics to sensing. Recent work has revealed the complex nature of these surface interactions for active particles. Experiments and simulations were summarized[257] that revealed the impact of charged nanoparticles upon the propulsion of an active colloid near to a boundary. The addition of charged nanoparticles not only decreased the average separation distance of a passive colloid because of depletion attraction, as expected, but also decreased the apparent propulsion tendency of a Janus colloid to near zero. Complementary agent-based simulations which considered the impact of hydrodynamics for active Janus colloids were conducted in a range of separation distances which was deduced from experiments. These simulations showed that the propulsion speed decreased monotonically with decreasing average separation distance. Although the trends found in experiments and simulations were in qualitative agreement, there was still a significant difference in the magnitude of the speed reduction. A quantitative difference was attributed to the influence of charged nanoparticles upon the conductivity of the active particle suspension. Follow-up experiments which delineated the impact of depletion and conductivity showed that both factors contributed to the reduction in speed for an active Janus particle. The experimental and simulated data suggested that it is necessary to consider the synergistic effects between various mechanisms which influence the interactions experienced by an active particle when near to a boundary.

The behavior of Janus particles fabricated from core silica particles decorated with gold nanoparticles on one hemisphere was studied[258] at an air/water interface. An unexpected reduction in the effective surface tension was observed in the presence of these chemically-modified Janus particles. Experiments which were conducted on the interfacial behavior of various control particles, including physically-modified Janus particles made from the same core silica particles coated with a thin gold layer, did not indicate the occurrence of significant surface tension effects. It was proposed that the chemical modification of particles in the form of a Janus structure was required in order to alter the surface tension, and the surfactant-like behavior of these particles was attributed to the presence of immersion forces.

The adhesion force and contact angle of gold-capped silica Janus particles and plain silica particles at an air/water interface were studied[259] via colloidal atomic force microscopy. Particles were attached to cantilevers at various orientations, and the wetting properties of the gold surface were varied by modification with dodecanethiol. This modification increased the hydrophobicity of the gold surface, and thus increased the difference between the contact angles of the gold hemisphere and the silica hemisphere and consequently increased the degree of amphiphilicity of the Janus particle. The colloidal probe was subsequently pushed into a stationary bubble from the water phase, followed by retraction back into the water phase. The Adhesion force was higher for Janus particles than for isotropic silica particles, regardless of the orientation of the anisotropic hemisphere. Particles with their polar half oriented towards the water, and with their apolar half facing the air, exhibited an increase in adhesion force and contact angle as the degree of amphiphilicity of the particles increased. No significant difference was observed, For particles of the reverse orientation, as the wetting properties changed. Both the adhesion force and the contact angle displayed an inverse relationship upon cap angle for particles having a higher degree of amphiphilicity. These results were of importance with regard to the use of Janus particles in stabilizing interfaces as well as for understanding the equilibrium height of Janus particles at the interface, as this would affect capillary interactions and hence self-assembly.

One can also envisage the case of moving boundaries. The situation has been considered[260] of a pair of identical microswimmers bounded by two harmonic traps in a thin sheared fluid film. Numerical investigation within the two-dimensional Oseen approximation suggested that the hydrodynamic pair coupling was long-ranged and was proportional to the ratio of the particle radius to the film thickness. When this ratio exceeded a certain value a transition occurred from a free regime, where each swimmer orbited within its own trap with a random phase, to a strongly synchronized regime wherein the two swimmers strongly repelled one another up to an average distance that was larger than both the trap distance and the free orbit diameter. The swimmers also tended to synchronize their positions opposite the center of the system. Beginning with a microscopic model for a spherically symmetrical active Janus particle, a study had previously been made[261] of the interaction between two such active swimmers. The fluid environment indeed mediated a long-range hydrodynamic interaction between them, and this interaction made both direct and indirect hydrodynamic contributions. The former contribution was due to fluid flow which originated from a moving swimmer and affected the motion of the other swimmer. The indirect contribution arose from the redistribution of ionic concentrations in the presence of both swimmers. Electrical forces which were exerted on the fluid by this ionic solution enhanced the flow pattern and thus altered the motion of both swimmers. A perturbation method for widely separated swimmers yielded analytical results for their translational and rotational dynamics. The overall modifications of the translational and rotational speeds of the swimmers scaled as third- and fourth-order functions of their separation, respectively.

The boundary-behaviour of axisymmetrical microswimming squirmers was analyzed theoretically for an inertialess Newtonian fluid at a no-slip interface[262]. Squirmers are often used as models for Janus particles, and an investigation was made here of the case of axisymmetrical tangential surface deformations, with a rotlet dipole being used to represent torque-motor swimmers such as flagellate bacteria. The resultant boundary dynamics could be reduced to a phase-plane involving the angle-of-attack and the distance from the boundary; using a simplifying time-reversal duality. Stable swimming adjacent to a no-slip boundary was reflected by the presence of stable fixed points. More generally, all types of fixed points as well as stable and unstable limit cycles occurred adjacent to a no-slip boundary following variations in the tangential deformations. There were constraints upon swimmer behavior in that swimmers which were characterized as pushers never exhibited stable limit cycles. All of the general aspects regarding no-slip boundaries were consistent with existing observations and geometrically faithful simulations. This suggested that the tangential squirmer was a relatively simple framework for making predictions of the complexities which are associated with axisymmetrical boundary swimming. In the presence of a free surface, with an asymptotically small capillary number and negligible leading-order surface deformations, no stable surface swimming was predicted to occur across the parameter space which was considered. This contrasted with experimental observations, but perhaps only in cases where surfactants might be present; thus possibly invalidating the assumption of a low capillary number. This implied that there is a need for surface deformation in order to ensure stable free-surface three-dimensional finite-size microswimming.

Applications

Janus particles in general have a wide range of applications where, in addition to those for which mobile particles are particularly useful, they are also exploited because of their more mundane static properties. For example, amphiphilic particles have been used to stabilize foam for drilling wells[263], for cell-tracking by magnetic particle imaging[264], for photo-chemotherapy[265] and for mechanochemical stimulation leading to cell death[266]. However, attention here is limited to mobile Janus particles with their two or more sides having differing surface chemical compositions or polarities, or two parts having differing geometrical morphologies. The rapid development of preparation methods for Janus particles, has caused attention to shift from their synthesis via selective surface modification, seeded crystallization, microfluidics, block copolymer self-assembly and electrochemical deposition, to their application in the fields of biomedicine, interfacial catalysis, surfactants, composite materials, micromotors and anti-fouling[267].

Catalytic Janus micromotors which incorporated Cd^{2+} or citrate ions have been used[268] as mobile microreactors for CdS quantum dot and gold nanoparticle synthesis. The motion of micromotor motion through μl-scale so-called reagent solutions results in the high-yield generation of the corresponding nanoparticles within the micromotor body, with negligible waste generation. Such nanoparticle generation can be attributed to the convective diffusion of reagents into the moving reactor. Control of the nanoparticle size and catalytic activity is possible by suitably choosing the motion of the microreactor. The use of confined reagents leads to efficient operation while using volumes of less than 800μl.

environmental remediation

Magnetocatalytic Janus micromotors consisting of phenylboronic acid-modified graphene quantum dots have been used[269] as ultra-fast sensors for the detection of bacterial endotoxins. The graphene quantum dots were loaded with platinum and iron oxide nanoparticles on one side. The different active regions permitted propulsion in the presence of hydrogen peroxide, or magnetic manipulation in the absence of chemical fuel. Fluorescence-quenching occurred when the graphene quantum dots encountered lipopolysaccharides in the target, and the phenylboronic acid tags acted as very specific recognition receptors. In another approach, micromotors were prepared[270] by using a Pickering emulsion approach to encapsulate the platinum nanoparticles required for bubble-propulsion plus receptor-functionalized quantum dots for selective binding with 3-deoxy-d-manno-oct-2-ulosonic acid. Lipopolysaccharides from *Salmonella enterica* were used as target endotoxins. Upon interacting with the quantum dots they caused a concentration-related fluorescence quenching. This reaction could detect concentrations as low as 0.07ng/ml of the endotoxin; which was below the level (275µg/ml) that is toxic to human beings. These micromotors could detect *Salmonella* toxin in food within 0.25h, rather than the several hours which were usually required.

Among the approaches, the use of Pickering emulsions with solid nanoparticles at the interface between two liquid phases was found[271] to be one of the most elegant methods for the synthesis of Janus colloidal particles having a controllable morphology over wide ranges of size and surface functionality. A monolayer of organic and inorganic nanoparticles can stabilize an emulsion droplet only when the hemispherical surface is chemically modified while the remaining surface is protected. This approach then offers the possibility of modifying the surface of nanoparticles with various functional groups, thus leading to Janus particles having complex structures.

Another approach to preparing Janus nanoparticles involves strawberry-like hierarchical composites having designed surface functional groups on both so-called satellites and a spherical core[272]. The "satellites" of the hierarchical composites can be freely varied, from iron oxide to silica nanoparticles. Results from transmission electron microscopy, Fourier transform infrared spectroscopy, and thermal gravimetric analysis measurements clearly prove the successful production of hybrid Janus silica nanoparticles coated by polystyrene and poly(acrylic acid). This technique demonstrates the vast flexibility of the abovementioned technique in terms of size, type, and surface chemistry design of Janus nanoparticles, which thus offers an additional approach to the current synthesis library of hybrid Janus nanoparticles.

So-called semi-decoration of silica particles with polydimethylamino ethylmethacrylate was carried out[273] by using surface-initiated atom transfer radical polymerization, based upon Pickering emulsion methods. Immobilized silica particles assembled at the interface of a paraffin-oil/water emulsion and a stable Pickering emulsion formed. Silver nanoparticles could then be absorbed onto the silica particles. The evaporation-driven self-assembly of colloidal-silica Janus particles has been evaluated[274]. Cyclodextrin- and azobenzene-modified samples were obtained by using the Pickering emulsion approach in which the particles were immobilized on solidified wax droplets and then functionalized. The particles were modified with 3-aminopropyl trimethoxysilane and afterwards reacted with tosyl-β-CD or phenylazo(benzoic acid), respectively. The preparation of hydrophobic/hydrophilic nanoparticles was studied[275] by synthesizing fluorinated Janus particles via the Pickering method and silica/wax emulsions. Partially wax-embedded silica particles were functionalized with an aminosilane. A so-called hairy hybrid Janus-type catalyst was prepared[276] which comprised an inorganic silica core covered with distinct, hydrophilic and hydrophobic, polymeric shells on opposite sides. A catalyst which consisted of silver or gold nanoparticles was immobilized on the hydrophilic polymer. Janus particles (200nm) were produced, which had poly(acrylic acid) as the hydrophilic polymer and polystyrene as the hydrophobic polymer, by using the Pickering emulsion method. These hairy Janus particles, furnished with very low amounts of immobilized silver or gold catalyst nanoparticles, could catalyze reactions efficiently. Titania-silica Janus particles which exhibited high photocatalytic activity were synthesized[277] by using a Pickering emulsion method and were calcined at one of two temperatures. Those which were calcined at 450C exhibited up to 14 times more adsorption and this was attributed to the electronic structure and its much higher average carrier lifetime of 1.98ns; as compared to the 0.18ns of other structures. Stearic-acid modified TiO_2 particles were prepared[278] via the impregnation approach and were used as a precursor to titania Janus particles. The results showed that titania Janus particles could be prepared via topo-selective surface modification. The stearic acid which was grafted onto the titania surface increased its hydrophobicity, encouraged charge separation and improved its adsorption of organic compounds.

A versatile large-scale synthesis method was described[279] for producing hybrid Janus nanoparticles having a silica core and a unilateral polymer corona. The method was based upon a modification of the Pickering emulsion polymerization technique, combined with surface-initiated atom transfer radical polymerization. In the first step, polyvinylacetate latex particles were prepared via Pickering emulsion polymerization. Colloidal stability was supplied by 30nm silica nanoparticles which adhered to the surface of growing polymer particles, giving a polymer latex which was armoured with a layer of tightly

Materials Research Forum LLC
https://doi.org/10.21741/9781644901199

immobilized nanoparticles; one side being immersed in the polymer particle. The exposed sides of the particles were then treated. Paramagnetic Fe_3O_4/SiO_2 Janus particles having phenyl and amino groups on opposite sides could be prepared[280] by using the Pickering method.

Polymeric Janus particles having sizes of 450 to 700nm were easily synthesized[281] by means of 1-step and 2-step one-pot emulsifier-free emulsion polymerization. Spherical and irregular-sphere morphologies which appeared during their evolution could be controlled by adjusting the amounts of hydrophilic or hydrophobic monomer. A surfactant-free approach was proposed[282] for the synthesis of non-spherical Janus particles with a temperature-dependent wettability on hydrophobic surfaces. Sub-micron particles consisting of a poly(styrene-co-divinylbenzene) core and a poly(N-isopropylacrylamide-co-methacrylic acid) shell were first prepared in order to stabilize styrene droplets in water; thus producing a Pickering emulsion. Upon heating to 80C and adding initiators to the aqueous phase, styrene droplets were polymerized and combined with the core–shell particles so as to form dumb-bell particles. Their shape could be controlled by adjusting the equilibration time of the Pickering emulsion at 80C.

A model system was proposed[283] for the scalable preparation of nanoscale Janus particles having dual-protein functionalization with ferritin and streptavidin. Silica nanoparticles (80nm) modified with azidosilane were used to prepare Pickering emulsions, with molten wax as the droplet phase. The azide-functionalized nanoparticles on the Pickering emulsion droplets were further subjected to face-selective silanization with biotin-polyethylene glycol ethoxy silane. Ferritin was then grafted onto the azide-functionalized side and the biotin groups were conjugated with streptavidin that was labelled with ultra-small gold nanoparticles.

Finally, Pickering emulsions have recently been used[284] to prepare semiconducting Janus particles by ultra-violet irradiation. Titanium dioxide particles that were pre-functionalized with an alkylsilane or a fluoroalkylsilane were used to obtain wax/water emulsion droplets with the particles partially embedded in the wax core. The emulsified wax droplets were then subjected to ultra-violet irradiation which photocatalytically degraded the silanes on the exposed part of the particle surfaces. This generated Janus particles with partial silane coatings. The particles were finally recovered by dissolving the wax. The Janus particles had differing wettabilities over their surface.

Returning now to the applications of mobile Janus particles, seawater has been used[285] as a fuel to propel Janus micromotors. The latter consisted of biodegradable and environmentally friendly magnesium microparticles which sported a nickel-gold bilayer patch for the purposes of magnetic control and surface modification. The seawater-driven

Materials Research Forum LLC
https://doi.org/10.21741/9781644901199

micromotors exploited macrogalvanic corrosion and chloride-pitting corrosion processes, thus eliminating any need for an external fuel supply.

A simple biotemplate method has been used[286] to produce platinum-free temperature-responsive micro-motors which exhibit very good recognition, capture and release abilities with regard to erythromycin in water. This method exploited surface imprint technology, the temperature-response of poly (N-isopropylacrylamide) hydrogels and the autonomous motion of Janus micromotors. The micromotors could be accurately guided, and later be recaptured, by using an external magnetic field.

Amphiphilic Janus particles which were covered with fluorinated chains were prepared, and their self-assembly and foam stabilization were studied over a limited range of concentrations[287]. Their structures and properties were characterized by using scanning electron microscopy, Fourier transform infra-red spectroscopy, fluorescence microscopy, pendant-drop tensiometry and foam scan. The adsorption of the particles at an interface saturated at a particle concentration of 0.6wt%, and an equilibrium surface tension of approximately 35mN/m was obtained. No further decrease in surface tension was observed with additional increases in particle concentration. At a concentration of 0.4wt%, Gibbs stability was achieved and the foam volume remained almost constant throughout the observation period. The liquid fraction of the foam also exhibited static equilibrium at each concentration, indicating that drainage was effectively arrested. The results confirmed that the use of amphiphilic Janus particles is an efficient approach to foam-stabilization.

Self-propelled Janus foam motors have been created[288], for the purpose of oil absorption on water, by loading a camphor/stearic-acid fuel mixture onto one end of a stearic acid-modified polyvinylalcohol foam. Such Janus foam motors exhibit extensive and efficient Marangoni-effect self-propulsion on water due to effective inhibition of any rapid release of the camphor by the hydrophobic stearic acid in the fuel mixture. The Marangoni-effect is exploited by certain insects as an escape mechanism[289]. The present simulacra-like motors could automatically search, capture and absorb oil droplets as they passed by and could then spontaneously self-assemble - following oil-absorption - due to both the self-propulsion of the motors and to the attractive capillary interaction between the motors and oil droplets. This made it easy to remove the motors from the water following treatment.

Iron-platinum bubble-propelled Janus micromotors which provided a high decontamination efficiency and efficient self-propulsion were prepared[290] via the asymmetrical deposition of catalytic platinum onto one hemisphere of zerovalent-iron microspheres. In the micromotor-peroxide system, the zerovalent iron acted as an heterogeneous Fenton-like catalyst for the degradation of organic pollutants while the hemispherical platinum layer, as usual, catalytically decomposed H_2O_2 into water and oxygen, thus providing bubble-propulsion at velocities of over 200μm/s in the presence of 5%H_2O_2. Complete oxidative degradation of methylene blue occurred in the presence of this 5%H_2O_2 solution after 1h of treatment, while zerovalent-iron microspheres removed only 12% of the methylene blue within 1h.

Capsules have been made by cross-linking aqueous microdroplets of chitosan biopolymer using glutaraldehyde[291]. Microfluidic tubing was used to generate chitosan droplets which contained nanoparticles having an iron core and a platinum shell. The droplets were then confined together with the cross-linking solution, and an external magnet pulled the nanoparticles together to form a collective patch on one end of each droplet. The resultant 150μm-diameter cross-linked capsules had an anisotropic structure and, when placed in H_2O_2 solution, the platinum shell catalyzed decomposition of the peroxide to form the oxygen bubbles which provided propulsion. The micromotors could be steered by an external magnet, due to the iron present in the nanoparticle. The micromotor could also adhere to an inert capsule, due to the soft nature of the two structures, and convey it to a desired location before release. The good antibacterial properties of chitosan were also combined[292] with the efficient water-powered mobility provided by magnesium micromotors. These Janus micromotors consisted of magnesium microparticles which were coated with biodegradable biocompatible polymers: poly(lactic-co-glycolic acid), alginate and chitosan. As a test, drinking water was contaminated with model *Escherichia coli* bacteria. This mobile-bactericide strategy produced a 27-fold improvement over the use of static chitosan-coated microparticles, with a 96% killing efficiency within 600s. Similar successes were had in treating seawater or fresh-water samples which were contaminated with unknown bacteria. Even before the advent of mobile Janus particles, chitosan and silver had been used[293] to form materials which exhibited a higher antimicrobial activity than did plain chitosan and which were effective against *Staphylococcus aureus, Bacillus subtilis, Escherichia coli* and *Salmonella choleraesuis* bacteria. A similar method[294] for killing *Escherichia coli* again exploited self-propelled Janus microbots which consisted of magnesium microparticles for propulsion and a magnetic iron layer for remote guidance. But here the bactericide was a silver-coated gold layer. The motion of the microbots again increased the probability of the silver nanoparticles contacting bacteria. Such contact provoked Ag^+ release in the cytoplasm, and the microbot motion increased the dispersion of the released ions. These silver-coated Janus microbots could kill more than 80% of the *Escherichia coli*, as compared with colloidal silver nanoparticles, which killed less than 35% in contaminated water solutions within 0.25h. The microbots could subsequently be recovered from the water by using a magnet.

A study was made[295] of catalytic Janus particles and *Escherichia coli* bacteria swimming in a two-dimensional colloidal crystal. The Janus particles orbited individual colloids and hopped stochastically between colloids at a hopping-rate which varied inversely with the hydrogen peroxide concentration. At high fuel concentrations, the orbits were stable for hundreds of revolutions and the orbital speed oscillated periodically as a result of hydrodynamic and perhaps phoretic interactions between the swimmer and 6 neighbouring colloids. Motile *E. coli* behaved very differently in the same colloidal crystal. That is, their circular orbits on plain glass were changed into long straight runs because the bacteria were unable to make turns within the crystal.

Bioluminescence imaging experiments have been used[296] to characterize the spatio-temporal patterns of bacterial self-organization in cultures of bioluminescent *Escherichia coli* and its mutants. Analysis of the effects of mutations showed that the spatio-temporal patterns which formed in standard dishes were not related to the chemotaxis of bacteria. These patterns were instead strongly dependent upon the properties of mutants, which characterized them as being self-phoretic - and not flagellar - swimmers. The observed patterns were essentially dependent upon the efficiency of proton-translocation across membranes and the smoothness of the cell surface. These characteristics were associated with surface activity and with phoretic mobility of a colloidal swimmer, respectively. Analysis of experimental data, together with mathematical modelling of the pattern formation, suggested that pattern-forming processes could be described by Keller–Segel like models of chemotaxis with logistic cell kinetics, that active cells could be viewed as being biochemical oscillators which exhibited phoretic drift and alignment and that spatio-temporal patterns in a suspension of growing *E. coli* bacteria formed due to phoretic interactions between oscillating cells which possess high metabolic activity.

Other self-propelled Janus particle micromotors have been based upon activated carbon[297]. They exhibited efficient movement under environmental conditions and promised the removal of a wide range of organic and inorganic pollutants. Their bubble-propulsion relied upon the asymmetrical deposition of a platinum patch onto the surface of an activated carbon microsphere. The rough surface of the latter produced an efficient catalytic microporous platinum layer, leading to copious bubble production and motor speeds of over 500µm/s. The combination of a high adsorption capacity due to the carbon, plus the rapid catalytic propulsive effect of the platinum, created in effect a highly efficient moving scoop which markedly accelerated water clean-up. The marked decontamination efficiency was applicable to heavy metals, nitro-aromatic explosives and organophosphorus nerve agents.

medical

There is a certain irony in the fact that Janus nanoparticles potentially have greatly beneficial uses in nanomedicine, while other nanoparticles in the environment are responsible for ill health. For drug delivery and medical diagnostics purposes, control of colloidal motion is a key factor in effective therapy. The autonomous motion of catalytic Janus colloids can remove the current need to induce and control colloid motion by using external fields, thus reducing the complexity of medical therapies and diagnostics. For materials applications which exploit colloidal self-assembly, the additional interactions introduced by catalytic activity and rapid motion are predicted to open up access to new reconfigurable and responsive structures[298]. In order to realize these goals, it is vital to develop methods to control both individual colloidal paths and collective behavior in motile catalytic colloidal systems. However, catalytic Janus colloids' trajectories are randomized by Brownian effects, and so require new strategies in order to be harnessed for transport. This is achievable by using a variety of different approaches. For example, self-assembly and control of catalyst geometry can introduce controlled amounts of rotary motion, or "spin" into chemical swimmer trajectories. The ability to control the degree of spin, or rotational velocity, of catalytic swimming bodies offers the prospect of increasing cargo binding-rates and achieving the theoretically proposed chiral diffusion behaviour. An assessment was made[299] of the possibility of imparting a well-defined spin to individual catalytic Janus swimmers by using glancing-angle metal evaporation onto a colloidal crystal with the aim of breaking the symmetry of the catalytic patch by shadowing with neighbouring colloids. This approach could demonstrate a well-defined relationship between the glancing angle and the ratio of the rotational to translational velocities. This permitted batches of colloids to be produced which had well-defined spin-rates ranging from 0.25 to 2.5Hz.

The use of active colloids for cargo transport has applications in fields ranging from targeted drug delivery to lab-on-a-chip systems. Janus particles, acting as mobile micro-electrodes, transport cargo which is trapped at the particle surface due to a dielectrophoretic mechanism. The cargo-loading properties of mobile Janus carriers over a wide range of frequencies and voltages could be compared[300] by expanding the frequency range of the carrier. The comparison revealed the effects of differing modes of carrier transport upon the loading capacity, and highlighted the differences between cargo that was trapped by positive or negative dielectrophoresis. It was shown that cargo-trapping resulted in a reduction in carrier velocities. This effect was more marked at low frequencies, where the cargo was trapped close to the substrate. There existed a maximum cargo loading capacity which decreased at large voltages, thus suggesting that there existed a strong interplay between trapping and hydrodynamic shear. It was demonstrated that control of the frequency permitted the existence of differing assemblies of binary colloidal solutions on a Janus particle. These results permitted the optimization of electrokinetic cargo transport and its selective application to a wide range of targets.

The excellent possibilities of environmental clean-up and biomedical treatment, offered by self-propelled nanorobots, have however been limited by the difficulty of producing them in industrial quantities. One solution is to develop[301] bubble-propelled FeO Janus nanomotors which could be prepared by using a simple solid-state thermal process. These nanomotors were stabilized by an ultra-thin iron -oxide shell and were propelled by their decomposition in citric acid and consequent asymmetrical bubble -production. The degradation of azo-dyes, for example, was markedly increased by using these moving FeO nanomotors, which acted as reducing agents.

Motion of biologically-compatible magnesium/platinum-poly(N-isopropylacrylamide) Janus micromotors in simulated body fluids or blood plasma was noted early on[302]. It was observed that pitting corrosion by chloride anions, and the buffering effect of simulated body fluids or blood plasma in removing the particle's passivation layer, were important in accelerating the reaction of magnesium with water and producing the hydrogen required for propulsion. The micromotors could take up, transport and release drug molecules by exploiting partially surface-attached thermoresponsive poly(N-isopropylacrylamide) hydrogel layers. These layers could easily be replaced by other polymers or antibodies. These micromotors were also anticipated to be useful for the detection and separation of heavy-metal ions, proteins, etc. Another magnesium-based Janus micromotor, powered by water, mimicked the behaviour of mobile cells[303]. It was constructed by incorporating red blood cell membranes, gold nanoparticles and alginate into the surface of magnesium microparticles that were partially embedded in Parafilm. The resultant micromotors exhibited guided propulsion in water without requiring any extra fuel, as well as propulsion in albumin-rich media. The red blood cell membrane coating endowed the Janus micromotors with the ability to absorb and neutralize both biological protein toxins and simulated nerve agents. This ability was of course greatly amplified by the marauder-like movement of the water-driven motors.

As well as imitating the lethality of white blood cells, it was found that Janus particles could also mimic the healing effects of cells. An artificial repair system was designed[304] in which patrolling self-propelled nanomotors sought out and localized themselves at microscopic cracks, thus satisfying the basic requirements needed for biological wound healing. The chemically-powered catalytic nanomotors consisted of conductive spherical gold-platinum particles which could autonomously detect and repair microscopic mechanical defects and restore the electrical conductivity of broken electrical pathways. The repair mechanism exploited the energetic wells and obstacles which were created by surface cracks, and which markedly affected the nanomotor dynamics and encouraged their localization at defects. The process was simulated for a range of particle speeds and densities and this confirmed that the nanomotors could form conductive patches which would repair scratched electrodes and restore conductive paths. Again like living cells, the micromotors could be caused to disappear from biological media when no longer required. The propulsive and self-destructive characteristics of magnesium-ZnO, magnesium-Si and zinc-iron Janus micromotors, together with those of single-component zinc micromotors, were determined[305]. The disappearance of the micromotors depended upon the various corrosion rates of the core-shell components.

Magnesium-gold micromotors have been used[306] as mobile chemical laboratories for the degradation into phenol of diphenyl phthalate, found in food and biological samples. The self-driven movement of micromotors through samples here resulted in the generation of hydrogen microbubbles and hydroxyl ions, which then degraded the diphenyl phthalate. The effectively-increased fluid transport both increased the sensitivity and lowered the detection level. Tests were fast ($\sim$300s) and nearly 100% accurate when applied to the direct analysis of diphenyl phthalate in food and biological samples. The magnesium-gold microspheres were also combined[307] with a self-assembled monolayer of meso-2,3-dimercaptosuccinic acid. The resultant micromotors again moved through environmental and biological media containing chlorides and surfactants, with no need for peroxide fuel or expensive platinum catalysts. They now acted as very efficient chelation centres which offered much shorter and more efficient water-cleaning than was possible when using static methods for treating zinc, cadmium and lead. Removal efficiencies of nearly 100% were reported for all of those metals following 120s treatment of serum, or brackish seawater, which had been doped with 500μg/l of each metal. The chelation mechanism was attributed to strong monolayer-type adsorption of metals by the meso-2,3-dimercaptosuccinic acid layer.

There is however a long history of Janus-particle preparation upon which to draw. Although that work was not aimed at producing self-propelling particles, it is worth recalling as a potential source of preparation strategies. Due to the non-centrosymmetric nature of Janus particles, their synthesis remained difficult for a long time[308].

A light-driven Janus micromotor for the removal of bacterial endotoxins and heavy metals has recently been described[309], with the biocompatible polycaprolactone polymer being used for the encapsulation of CdTe or CdSe/ZnS quantum dots as photo-active materials and an asymmetrical Fe_3O_4 patch providing propulsion. The micromotors could be activated by visible light (470–490nm) to trigger propulsion in peroxide or glucose media via diffusiophoresis. Efficient propulsion was observed in complex samples such as human blood serum, and efficient endotoxin removal was demonstrated by using lipopolysaccharides from *Escherichia coli* as a model toxin. These micromotors could also be used for mercury removal by cationic exchange with the CdSe/ZnS core–shell quantum dots.

It can be desirable to increase the motility of nano-scale or microscale swimmers, such as self-propelled Janus particles, as agents of chemical reactions or carriers of weak sperm cells for better chances of fertilization. This is based upon the concept that motility can be transferred from a more-active guest to a less-active host. Numerical simulations of motility transfer were performed[310] for two typical cases: for interacting particles having a weak inertia effect by analyzing the velocity distributions, and for interacting overdamped particles by studying the effusion rate. In both cases, motility transfer was detected with motility enhancement of the host species by a factor of up to 4.

Janus particles have found some use with regard to virus activity, and it was suggested[311] that magnetic Janus microparticles could be used to control the stimulation of T-cell signalling; with single-cell precision. Janus particles were designed which were magnetically responsive on one hemisphere and which could stimulate T-cell activity using the other hemisphere. By controlling the rotation and motion of Janus particles via an external magnetic field, it was possible to manipulate the orientation of particle–cell recognition and thus the initiation of T-cell activation. In earlier work[312], bifunctional Janus particles had been engineered in which the spatial distributions of two ligands, anti-CD3 and fibronectin, had been made to mimic the so-called bull's-eye protein pattern which formed in the membrane junction between a T-cell and an antigen-bearing cell. Various levels of T-cell activation could be chosen simply by switching the spatial distributions of the two ligands on the particle surface. The ligand pattern also affected the clustering of intracellular proteins. Migration away from self-produced chemicals, known as chemorepulsion, generates a generic route to clustering and pattern-formation among self-propelled colloids[313]. The clustering instability can be caused by anisotropic chemical production or by a delayed orientational response to changes in the chemical environment. In either case, chemorepulsion creates clusters of self-limiting area which grow linearly with self-propulsion speed. This agreed with observations of dynamic clusters in Janus colloids.

Mesoporous carbon nanoparticles, like those described previously, have been prepared[314] by depositing platinum nanoparticles onto half of the carbon spheres. These were capable of carrying 1370mg/g of the anticancer drug, doxorubicin hydrochloride. An increase in velocity occurred upon illumination with near-infra-red light, due to local thermophoresis. An increased velocity could also result from adding H_2O_2, which then propelled the particles due to the usual asymmetrical decomposition. An increased velocity also improved the efficiency of drug delivery because it increased the number of hollow mesoporous carbon nanospheres which adhered to the surface of a cancer cell. Another new drug-delivery system[315] has been based upon poly(lactic-co-glycolic acid) magnetic Janus carrier particles which exploit electrohydrodynamic co-jetting. The drug-carrying Janus particles could be loaded with, for example, paclitaxel for killing cancer cells, Fe_3O_4 nanoparticles for target location and rhodamine-B for fluorescence-tracing. The structure of the drug-loaded magnetic Janus particles permitted higher drug loads to be carried, and also increased the cumulative release-rate of those drugs in various media. These drug-loaded magnetic Janus particles exhibited a high toxicity towards only human-breast cancer cells. The particles were also more lethal to cancer cells than were high-concentration paclitaxel suspensions. When guided using an external magnetic field, the drug-loaded magnetic Janus particles could easily target, and accumulate at, cancer cells. Yet another type of Janus nanocarrier was created[316] by sputtering gold onto mesoporous silica particles, and was then propelled by using near-infrared light. This nanocarrier could deeply penetrate a tumour via the thermomechanical percolation of cytomembranes and subsequent controlled drug release. This approach improved the use of nanocarriers and reduced the side-effects which were caused by the drugs being carried. In general, the movement of drug-loaded nanocarriers is impeded by various physiological and pathological barriers involving blood circulation and tissue penetration. It has very recently been suggested[317] that self-propelled fibre-rod micromotors could overcome transportation barriers to chemotherapeutic drugs. Fibre-rods having a distinct Janus structure were prepared via side-by-side electrospinning and urease and folate were attached to the sides of the rods as a power-source and targeting ligand. The density of the urease coating determined the velocity and trajectory of the particles, and high velocities and high mean-square displacements were observed in both phosphate-buffered saline and simulated extracellular tumour-tissue. In spite of a lack of interference with the up-take pathway, self-propelled motion promoted folate-mediated cell internalization of the particles and maximized the poisoning and apoptosis rate of tumour cells. The self-propulsion force arising from one side of the rods increased lateral drift and extravasation across blood-vessel walls. This increased their accumulation, in the tumour tissue, by a factor of about 2 and increased doxorubicin accumulation by a factor of about 2.6. Treatment with these particles markedly inhibited tumour growth, prolonged survival rate and provoked marked tumour necrosis without introducing histopathological or hematological abnormalities into normal tissue.

A novel polymer-based motor was prepared[318] by means of template-assisted polyelectrolyte layer-by-layer deposition of a thin gold layer onto one side, followed by chemical immobilization of a catalytic enzyme. These Janus capsule motors could self-propel at body-temperature in the presence of 0.1% peroxide fuel and exhibited a higher speed than that of platinum-based motors. They were used for encapsulation of the anticancer drug, doxorubicin, so as to be able to steer it towards a target cell layer by using an external magnetic field, whereupon drug release was triggered by using near-infrared light. Janus-like nanohybrid particles were prepared[319] in which an internal heterojunction, composed of inorganic gold stars and organic bis-pyrene nano-aggregates, was coated with a pegylated silica shell. The nanohybrids exhibited thermophoresis under near-infrared irradiation. A series of nanohybrids was prepared which had a constant concentration of gold stars but varying concentrations of mercapto-propyl triethoxysilane and bis-pyrene. The nanohybrids exhibited controlled motion under near-infrared laser control. They possessed a higher cell-killing propensity than did bis-pyrene-free particles at the same concentration. The active motion of the nanohybrids was thought to increase the temperature of treated cells by converting kinetic energy into thermal energy; again killing cancer cells. The nanohybrids also exhibited enhanced photothermal treatment *in vivo*, and killed tumour cells 2 days after treatment via photothermal ablation using near-infrared laser irradiation. A reported near-infrared light-powered mesoporous silica Janus nanomotor with macrophage cell membrane cloaking could actively seek out cancer cells and thermomechanically perforate the cell membrane[320]. Under exposure to near-infrared light, a heat gradient was created across the Janus boundary of the mesoporous silica nanomotor due to the photothermal effect of the gold half-shells. This resulted in a self-thermophoretic force which propelled the silica nanomotor. In a biological medium, macrophage cell membrane camouflaging not only prevented dissociative biological blocks from sticking to the Janus nanomotor but also improved the search-sensitivity of the nanomotors by specifically recognizing cancer cells. These propulsive and recognition abilities permitted the Janus mesoporous silica nanomotor to find and bind itself to the membranes of cancer cells. Interactions between Janus particles and membranes were investigated[321] by using dissipative particle dynamics, and this revealed that there were two different interaction-modes: insertion and engulfment. The initial orientation and properties of the particles here had a marked effect upon the interactions. That is, when the hydrophilic part of the particle was close to the membrane, or when the particle had a larger cross-sectional area and greater hydrophilic coverage, the particle was more likely to be engulfed. In the case of membranes containing lipid rafts, the Janus particle could easily detach itself from a membrane after it was engulfed by the raft.

Silver–silica Janus particles, with amine, thiol and epoxy on the exposed surface of SiO_2 particles have been synthesized[322] and tested for antimicrobial activity. Due to their easy dispersibility, samples having silver nanoparticles with a diameter of about 3nm exhibited a much lower minimum bactericidal concentration when compared with that of conventional isotropic silver nanoparticles having a similar diameter. Isotropic silver nanoparticles and functionalized $Ag-SiO_2$ Janus particles were attached to cotton fabric by using the so-called exhaustion method, followed by curing. These fabrics were then tested for antimicrobial activity and washability. The $Ag-SiO_2$ Janus particles, due to the presence of various functional groups on half of the surface, could be attached to cellulosic substrates and thus impart a lasting antimicrobial activity. Such Janus particles are perfect for mimicking natural anisotropic architectures and directional interactions. A simple combination of seeded-emulsion polymerization and thiol chemistry has been developed[323] for the synthesis of particles having glucose molecules on one side. Such biofunctional Janus particles exhibit a region-selective binding to protein and thus a considerable biomimicry which could allow the particles to deliver drugs. Fluorescence microscopy confirmed that these particles indeed exhibited a region-selective binding to proteins.

As an astonishing example of the value of Janus particles in the field of medical research, it is to be noted that self-thermophoretically driven gold-silica particles can be prepared which can simultaneously stretch and partially melt a single double-stranded DNA molecule[324]. The force which acts upon the molecule is in the pN-range and this can be used to monitor the entropic stretching regime of the DNA molecule, while local temperature increases on the gold side of the particle can provoke partial DNA dehybridization. Another form of active droplet was proposed[325] which was attracted or repelled by certain obstacle geometries. Such droplets consisted of a water/ethanol mixture dispersed in an oil/surfactant solution. Due to mass exchange between the fluid phases during self-propulsion, the originally homogeneous droplets could spontaneously de-mix and evolve into Janus droplets. Molecules such as DNA could separate into the trailing ethanol-rich droplet and be carried to their target location. The delayed onset of phase separation could provide a means for controlling the delivery time, while long-range hydrodynamic interactions and short-range wetting forces could also be exploited.

Equally astonishing is the use of these particles to treat gross injuries in addition to infections. Such an example is the rapid and secure control of perforating and irregular haemorrhages[326]. That is, bleeding wounds having irregular external and internal wound shapes and located deeply within complex and hidden sites. Current hemostatic materials do not effectively control haemorrhage because of their limited ability to access the bleeding source or to coagulate blood. Janus self-propelled hemostatic particles were prepared by the uniaxial growth of flower-like calcium carbonate crystals onto negatively-modified microporous starch. The as-synthesized hemostatic particles were then loaded with thrombin and were powered by the $CaCO_3$ component and protonated tranexamic acid. The particles were capable of moving contrary to the blood flow, thus permitting them to reach deep bleeding sites and induce a synergistic blood coagulation which efficiently halted the haemorrhaging. The self-propelling Janus hemostatic particles were sufficiently available in the case of deep bleeding sites in a liver or a femoral artery; with the haemorrhage being controlled within some 50s and 180s, respectively.

It nevertheless has to be recalled that many microswimmers for medical drug and cargo delivery rely upon toxic fuel sources and component materials. Bacteria are an excellent motor alternative because they are powered by using the surrounding biological fluid. Odd as it seems, the notorious *Escherichia coli* bacterium, frequently mentioned here in contexts such as environmental contamination, have been integrated into metal-capped polystyrene Janus particles[327]. Fabrication of such biohybrids is rapid and simple for a microswimmer which is capable of magnetic guidance and which can transport an anti-cancer agent. Cell-adhesion is simple because the *E. coli* adheres only to the metal cap of the particle; thus leaving the polystyrene surface available for drug attachment. Bacterium adhesion has been investigated for platinum, iron, titanium or gold caps; revealing a marked preference for platinum surfaces.

Rotary motion combined with gravity, produces well-defined orientated helical trajectories. In addition, when catalytic colloids interact with topographical features, such as edges and trenches, they are steered. This gives rise to a new approach for autonomous colloidal microfluidic transport that could be deployed in future lab-on-a-chip devices. Chemical gradients can also influence the motion of catalytic Janus colloids, for example, to cause collective accumulations at specific locations. However, at present, the predicted theoretical degree of control over this phenomenon has not been fully verified in experimental systems. Collective behavior control for chemical swimmers is also possible by exploiting the potential for the complex interactions in these systems to allow access to self-assembled, dynamic and reconfigurable ordered structures. Again, current experiments have not yet accessed the breadth of possible behavior. Consequently, continued efforts are required to understand and control these interaction mechanisms in real-world systems. Ultimately, this will help realize the use of catalytic Janus colloids for tasks that require well-controlled motion and structural organization, enabling functions such as analyte capture and concentration, or targeted drug delivery.

Micromotor-based biosensing studies have been used[328] to functionalize the surface of micromotors having specific molecular probes for the binding of a target analyte, thus limiting the use of the micromotor to that specific target. The new concept was here introduced of using a non-functionalized micromotor as a generic cargo-carrier which was able to perform the dynamic loading, transport and release of functionalized beads. This approach permitted the use of a given micromotor system for sensing various targets using functionalized beads, and demonstrated the use of micromotors as a versatile means for biosensing. A simplified microfluidic design was introduced which could be used for immunosensing or DNA binding tests without requiring complicated fluid-handling.

electronics

Janus particles have attracted much attention due to their potential use in electronic applications[329]. For the application of Janus particles to high-resolution displays, and as light sources for optical circuits and fluorescent probes, the particles should be nanosized in order to ensure high-resolution display characteristics, responsiveness to external stimuli and high fluorescence. It remains a challenge however to develop such highly fluorescent nanoscale Janus particles and control their alignment. Magnetoresponsive Janus particles, whose orientation can be controlled by an external magnetic field, are easily prepared by the introduction of polymer-coated magnetic nanoparticles into the hemispheres of Janus particles. If these magnetoresponsive Janus particles can be combined with a strongly fluorescing system, then they can be ideal candidates for the components of various applications. As one example, Janus particles have been prepared which incorporated a fluorescent dye and gold nanoparticles on one side. The optical properties of such particles were assessed with regard to the response of composite Janus particles containing dyes to an external magnetic field.

Janus particles possessing a double functionality have been synthesized[330] via the asymmetrical loading of alginate hydrogel beads with Prussian Blue thus producing microswimmers which exhibited an oscillatory behavior coupled to chemical light emission. This resulted from the combination of Prussian Blue acting as a catalyst and permitting concomitant light emission and oxygen production in the presence of luminol and hydrogen peroxide, together with a differential porosity which led to the asymmetrical release of the oxygen bubbles which propelled the particles. Synthesis of these materials was possible by using an electric field-based symmetry-breaking with ionically cross-linked alginate beads.

The information deduced from their translational dynamics has made colloidal particles invaluable micro-sized probes in analytical techniques such as particle-imaging velocimetry. Janus particles, with their partial shadowing by metal caps, appear to be directional when imaged by means of optical microscopy. Detectable rotational dynamics are also introduced into otherwise spherically symmetrical systems. Two novel image-analysis methods have been proposed[331] which identify the body-center of gold-capped Janus particles and distinguish between particle orientations. One method tracks the optical centroid, determines the in-plane tilt-angle and locates the body center. The other method tracks the diffraction-ring center and the moon-phase center in order to reveal the

Materials Research Forum LLC
https://doi.org/10.21741/9781644901199

in-plane orientation. Both methods can identify body-centers and orientations with a resolution of 0.5 pixels and an angular uncertainty of less than 10°, making it possible to probe the local medium response at sub-micron length-scales. Such methods were used to track dilute suspensions of gold-capped Janus particles which were undergoing Brownian motion in viscous glycerol-water mixtures. The mean-square displacements and angular mean-square displacements which were deduced from experiments could be related to the viscosity of the medium in terms of Stokes-Einstein and Perrin rotational diffusion, respectively. The viscosities which were calculated on the basis of the diffusion coefficients, inferred from the translational and rotational dynamics of microprobes, differed by less than 6%. This discrepancy was attributed to wall and gravitational effects.

Self-propelling micromotors are increasingly promising micro-scale and nano-scale tools for single-cell analysis. The field gradients which are required to manipulate matter by dielectrophoresis can be induced at the surface of a self-propelling metallodielectric Janus particle under an externally applied electric field. It thus effectively acts as a mobile floating microelectrode. It has been demonstrated[332] that the application of an external electric field can trap and transport bacteria and, moreover, selectively electroporate the trapped bacteria. Selective electroporation, made possible by the local intensification of the electric field induced by the particle, was induced under continuous alternating-current and pulsed-signal conditions. This approach is applicable to bacteria and Janus particles, and to a wide range of cell-types and micromotor designs. It is therefore a valuable experimental tool for single-cell analysis and targeted delivery.

Keyword Index

Self-Propelled Janus Particles   Materials Research Forum LLC
Materials Research Foundations **93** (2021)   https://doi.org/10.21741/9781644901199

About the Author

Dr. Fisher has wide knowledge and experience of the fields of engineering, metallurgy and solid-state physics, beginning with work at Rolls-Royce Aero Engines on turbine-blade research, related to the Concord supersonic passenger-aircraft project, which led to a BSc degree (1971) from the University of Wales. This was followed by theoretical and experimental work on the directional solidification of eutectic alloys having the ultimate aim of developing composite turbine blades. This work led to a doctoral degree (1978) from the Swiss Federal Institute of Technology (Lausanne). He then acted for many years as an editor of various academic journals, in particular *Defect and Diffusion Forum*. In recent years he has specialised in writing monographs which introduce readers to the most rapidly developing ideas in the fields of engineering, metallurgy and solid-state physics. He is co-author of the widely-cited student textbook, *Fundamentals of Solidification*. Google Scholar credits him with 7228 citations and a lifetime h-index of 12.

References

[1] Alarcón-Correa, M., Walker, D., Qiu, T., Fischer, P., European Physical Journal - Special Topics, 225[11-12] 2016, 2241-2254. https://doi.org/10.1140/epjst/e2016-60067-1

[2] Ishimoto, K., Yamada, M., SIAM Journal on Applied Mathematics, 72[5] 2012, 1686-1694. https://doi.org/10.1137/110853297

[3] Purcell, E.M., American Journal of Physics, 45[1] 1977, 3-11. https://doi.org/10.1119/1.10903

[4] Lugli, F., Brini, E., Zerbetto, F., Journal of Physical Chemistry C, 116[1] 2012, 592-598. https://doi.org/10.1021/jp205018u

[5] Mach, E., The Science of Mechanics - a Critical and Historical Account of its Development, 6th edition, Open Court, 1960, 388-390.

[6] Dijkink, R.J., van der Dennen, J.P., Ohl, C.D., Prosperetti, A., Journal of Micromechanics and Microengineering, 16, 2006, 1653–1659. https://doi.org/10.1088/0960-1317/16/8/029

[7] Seife, C., Science, 299, 2003, 1295. https://doi.org/10.1126/science.299.5611.1295a

[8] Michelin, S., Lauga, E., Scientific Reports, 7, 2017, 42264. https://doi.org/10.1038/srep42264

[9] Ebbens, S., Sadeghi, A., Howse, J., Golestanian, R., Jones, R., Materials Research Society Symposium Proceedings, 1346, 2011, 49-52. https://doi.org/10.1557/opl.2011.1003

[10] Michelin, S., Lauga, E., Journal of Fluid Mechanics, 747, 2014, 572-604. https://doi.org/10.1017/jfm.2014.158

[11] Tătulea-Codrean, M., Lauga, E., Journal of Fluid Mechanics, 856, 2018, 921-957. https://doi.org/10.1017/jfm.2018.718

[12] Tomlinson, C., Experimental Essays, Virtue Brothers & Co., 1863, 7-60.

[13] Frankel, A.E., Khair, A.S., Physical Review E, 90[1] 2014, 013030. https://doi.org/10.1103/PhysRevE.90.013030

[14] Crowdy, D.G., Journal of Fluid Mechanics, 735, 2013, 473-498. https://doi.org/10.1017/jfm.2013.510

[15] Wu, M., Zhang, H., Zheng, X., Cui, H., AIP Advances, 4[3] 2014, 031326. https://doi.org/10.1063/1.4868375

[16] ten Hagen, B., van Teeffelen, S., Löwen, H., Condensed Matter Physics, 12[4] 2009, 725-738. https://doi.org/10.5488/CMP.12.4.725

[17] Mozaffari, A., Sharifi-Mood, N., Koplik, J., Maldarelli, C., Physics of Fluids, 28[5] 2016, 053107. https://doi.org/10.1063/1.4948398

[18] Speck, T., Physical Review E, 99[6] 2019, 060602. https://doi.org/10.1103/PhysRevE.99.060602

[19] Brosseau, Q., Usabiaga, F.B., Lushi, E., Wu, Y., Ristroph, L., Zhang, J., Ward, M., Shelley, M.J., Physical Review Letters, 123[17] 2019, 178004. https://doi.org/10.1103/PhysRevLett.123.178004

[20] Zheng, X., Wu, M., Kong, F., Cui, H., Silber-Li, Z., Journal of Visualization, 18[3] 2015, 425-435. https://doi.org/10.1007/s12650-015-0299-5

[21] Cui, H., Tan, X., Zhang, H., Nanotechnology and Precision Engineering, 12[5] 2014, 340-345.

[22] Wu, M.L., Zheng, X., Cui, H.H., Li, Z.H., Chinese Journal of Hydrodynamics A, 29[3] 2014, 274-281.

[23] Gao, W., Pei, A., Dong, R., Wang, J., Journal of the American Chemical Society, 136[6] 2014, 2276-2279. https://doi.org/10.1021/ja413002e

[24] Ma, X., Hahn, K., Sanchez, S., Journal of the American Chemical Society, 137[15] 2015, 4976-4979. https://doi.org/10.1021/jacs.5b02700

[25] Zhang, Q., Dong, R., Chang, X., Ren, B., Tong, Z., ACS Applied Materials and Interfaces, 7[44] 2015, 24585-24591. https://doi.org/10.1021/acsami.5b06448

[26] Chatterjee, P., Tang, E.M., Karande, P., Underhill, P.T., Physical Review Fluids, 3[1] 2018, 014101. https://doi.org/10.1103/PhysRevFluids.3.014101

[27] Saad, S., Natale, G., Soft Matter, 15[48] 2019, 9909-9919. https://doi.org/10.1039/C9SM01801H

[28] Ebbens, S., Tu, M.H., Howse, J.R., Golestanian, R., Physical Review E, 85[2] 2012, 020401. https://doi.org/10.1103/PhysRevE.85.020401

[29] Brown, A., Poon, W., Soft Matter, 10[22] 2014, 4016-4027. https://doi.org/10.1039/C4SM00340C

[30] Ebbens, S., Gregory, D.A., Dunderdale, G., Howse, J.R., Ibrahim, Y., Liverpool, T.B., Golestanian, R., EPL, 106[5] 2014, 58003. https://doi.org/10.1209/0295-5075/106/58003

[31] Shemi, O., Solomon, M.J., Journal of Physical Chemistry B, 122[44] 2018, 10247-10255. https://doi.org/10.1021/acs.jpcb.8b08303

[32] Xing, Y., Pan, Q., Du, X., Xu, T., He, Y., Zhang, X., ACS Applied Materials and Interfaces, 11[10] 2019, 10426-10433. https://doi.org/10.1021/acsami.8b22612

[33] Longbottom, B.W., Bon, S.A.F., Scientific Reports, 8[1] 2018, 4622. https://doi.org/10.1038/s41598-018-22917-2

[34] Choudhury, U., Soler, L., Gibbs, J.G., Sanchez, S., Fischer, P., Chemical Communications, 51[41] 2015, 8660-8663. https://doi.org/10.1039/C5CC01607J

[35] Ye, S., Carroll, R.L., ACS Applied Materials and Interfaces, 2[3] 2010, 616-620. https://doi.org/10.1021/am900839w

[36] Moran, J.L., Posner, J.D., Journal of Fluid Mechanics, 680, 2011, 31-66. https://doi.org/10.1017/jfm.2011.132

[37] Moran, J.L., Posner, J.D., Physics of Fluids, 26[4] 2014, 042001. https://doi.org/10.1063/1.4869328

[38] Rao, D.V., Reddy, N., Fransaer, J., Clasen, C., Journal of Physics D, 52[1] 2019, 014002. https://doi.org/10.1088/1361-6463/aae6f6

[39] Reddy, N.K., Palangetic, L., Stappers, L., Buitenhuis, J., Fransaer, J., Clasen, C., Journal of Materials Chemistry C, 1[23] 2013, 3646-3650. https://doi.org/10.1039/c3tc30176a

[40] Zhang, Y., Cui, H.H., Zhou, M., Chen, L., Chinese Journal of Hydrodynamics A, 32[2] 2017, 237-246.

[41] Yan, W., Brady, J.F., Journal of Chemical Physics, 145[13] 2016, 134902. https://doi.org/10.1063/1.4963722

[42] Würger, A., Physical Review Letters, 115[18] 2015, 188304. https://doi.org/10.1103/PhysRevLett.115.188304

[43] Alarcon, F., Navarro-Argemí, E., Valeriani, C., Pagonabarraga, I., Physical Review E, 99[6] 2019, 062602. https://doi.org/10.1103/PhysRevE.99.062602

[44] Shen, Z., Würger, A., Lintuvuori, J.S., European Physical Journal E, 41[3] 2018, 39. https://doi.org/10.1140/epje/i2018-11649-0

[45] Ishikawa, T., Micromachines, 10[1] 2019, 33. https://doi.org/10.3390/mi10010033

[46] Kuron, M., Kreissl, P., Holm, C., Accounts of Chemical Research, 51[12] 2018, 2998-3005. https://doi.org/10.1021/acs.accounts.8b00285

[47] Zhou, G.Y., Chen, L., Zhang, H.Y., Cui, H.H., Acta Physica Sinica, 66[8] 2017, 084703.

[48] Zhou, C., Zhang, H.P., Tang, J., Wang, W., Langmuir, 34[10] 2018, 3289-3295. https://doi.org/10.1021/acs.langmuir.7b04301

[49] Singh, D.P., Uspal, W.E., Popescu, M.N., Wilson, L.G., Fischer, P., Advanced Functional Materials, 28[25] 2018, 1706660. https://doi.org/10.1002/adfm.201706660

[50] Moyses, H., Palacci, J., Sacanna, S., Grier, D.G., Soft Matter, 12[30] 2016, 6357-6364. https://doi.org/10.1039/C6SM01163B

[51] Miloh, T., Nagler, J., Electrophoresis, 39[19] 2018, 2417-2424. https://doi.org/10.1002/elps.201800211

[52] Bregulla, A.P., Cichos, F., Journal of Chemical Physics, 151[4] 2019, 044706. https://doi.org/10.1063/1.5088131

[53] Bregulla, A.P., Cichos, F., Proceedings of SPIE, 9922, 2016, 99221L.

[54] Baraban, L., Streubel, R., Makarov, D., Han, L., Karnaushenko, D., Schmidt, O.G., Cuniberti, G., ACS Nano, 7[2] 2013, 1360-1367. https://doi.org/10.1021/nn305726m

[55] Jiang, H.R., Yoshinaga, N., Sano, M., Physical Review Letters, 105[26] 2010, 268302. https://doi.org/10.1103/PhysRevLett.105.268302

[56] Chen, Y.L., Yang, C.X., Jiang, H.R., Scientific Reports, 8[1] 2018, 5945.

[57] Gaspard, P., Kapral, R., Journal of Statistical Mechanics - Theory and Experiment, 7, 2019, 074001. https://doi.org/10.1088/1742-5468/ab252f

[58] Baier, T., Tiwari, S., Shrestha, S., Klar, A., Hardt, S., Physical Review Fluids, 3[9] 2018, 094201. https://doi.org/10.1103/PhysRevFluids.3.094202

[59] De Graaf, J., Samin, S., Soft Matter, 15[36] 2019, 7219-7236. https://doi.org/10.1039/C9SM00886A

[60] Xing, Y., Du, X., Xu, T., Zhang, X., Soft Matter, 16[41] 2020, 9553-9558. https://doi.org/10.1039/D0SM01355B

[61] Heidari, M., Bregulla, A., Landin, S.M., Cichos, F., Von Klitzing, R., Langmuir, 36[27] 2020, 7775-7780. https://doi.org/10.1021/acs.langmuir.0c00461

[62] Gangwal, S., Cayre, O.J., Velev, O.D., Langmuir, 24[23] 2008, 13312-13320. https://doi.org/10.1021/la8015222

[63] Zhang, L., Zhu, Y., Applied Physics Letters, 96[14] 2010, 141902. https://doi.org/10.1063/1.3378687

[64] Honegger, T., Lecarme, O., Berton, K., Peyrade, D., Journal of Vacuum Science and Technology B, 28[6] 2010, C6I14-C6I19. https://doi.org/10.1116/1.3502670

[65] Honegger, T., Lecarme, O., Berton, K., Peyrade, D., Microelectronic Engineering, 87[5-8] 2010, 756-759. https://doi.org/10.1016/j.mee.2009.11.145

[66] Boymelgreen, A.M., Miloh, T., Physics of Fluids, 23[7] 2011, 072007. https://doi.org/10.1063/1.3609804

[67] Boymelgreen, A.M., Miloh, T., Physics of Fluids, 24[8] 2012, 082003. https://doi.org/10.1063/1.4739932

[68] Boymelgreen, A.M., Miloh, T., Electrophoresis, 33[5] 2012, 870-879.
https://doi.org/10.1002/elps.201100446

[69] Chen, J., Zhang, H., Zheng, X., Cui, H., AIP Advances, 4[3] 2014, 031325.
https://doi.org/10.1063/1.4868373

[70] Lee, T.C., Alarcón-Correa, M., Miksch, C., Hahn, K., Gibbs, J.G., Fischer, P., Nano
Letters, 14[5] 2014, 2407-2412. https://doi.org/10.1021/nl500068n

[71] Molotilin, T.Y., Lobaskin, V., Vinogradova, O.I., Journal of Chemical Physics,
145[24] 2016, 244704. https://doi.org/10.1063/1.4972522

[72] Dou, Y., Cartier, C.A., Fei, W., Pandey, S., Razavi, S., Kretzschmar, I., Bishop,
K.J.M., Langmuir, 32[49] 2016, 13167-13173.
https://doi.org/10.1021/acs.langmuir.6b03361

[73] Ibrahim, Y., Golestanian, R., Liverpool, T.B., Journal of Fluid Mechanics, 828, 2017,
318-352. https://doi.org/10.1017/jfm.2017.502

[74] Zhao, K., Li, D., Journal of Micromechanics and Microengineering, 27[9] 2017,
095007. https://doi.org/10.1088/1361-6439/aa7eae

[75] Lin, C.H., Chen, Y.L., Jiang, H.R., RSC Advances, 7[73] 2017, 46118-46123.
https://doi.org/10.1039/C7RA08527C

[76] Brown, A.T., Poon, W.C.K., Holm, C., De Graaf, J., Soft Matter, 13[6] 2017, 1200-
1222. https://doi.org/10.1039/C6SM01867J

[77] Bayati, P., Najafi, A., Journal of Chemical Physics, 150[23] 2019, 234902.
https://doi.org/10.1063/1.5101023

[78] Croze, O.A., Martinez, V.A., Jakuszeit, T., Dellarciprete, D., Poon, W.C.K., Bees,
M.A., New Journal of Physics, 21[6] 2019, 063012. https://doi.org/10.1088/1367-
2630/ab241f

[79] Zhou, T., Ji, X., Shi, L., Zhang, X., Deng, Y., Joo, S.W., Electrophoresis, 40[6] 2019,
993-999. https://doi.org/10.1002/elps.201800368

[80] Shen, C., Jiang, Z., Li, L., Gilchrist, J.F., Ou-Yang, H.D., Micromachines, 11[3] 2020,
334. https://doi.org/10.3390/mi11030334

[81] Wang, L.L., Cui, H.H., Zhang, J., Zheng, X., Wang, L., Chen, L., Acta Physica Sinica,
65[22] 2016, 220201.

[82] Zheng, X., Cui, H., Li, Z., Chinese Science Bulletin, 62[2-3] 2017, 167-185.
https://doi.org/10.1360/N972016-01139

[83] Pourrahimi, A.M., Villa, K., Ying, Y., Sofer, Z., Pumera, M., ACS Applied Materials
and Interfaces, 10[49] 2018, 42688-42697. https://doi.org/10.1021/acsami.8b16217

[84] Yuan, Y., Gao, C., Wang, D., Zhou, C., Zhu, B., He, Q., Beilstein Journal of Nanotechnology, 10, 2019, 1324-1331. https://doi.org/10.3762/bjnano.10.131

[85] María-Hormigos, R., Escarpa, A., Goudeau, B., Ravaine, V., Perro, A., Kuhn, A., Advanced Materials Interfaces, 7[6] 2020, 1902094. https://doi.org/10.1002/admi.201902094

[86] Hwang, J., Yang, H.M., Lee, K.W., Jung, Y.I., Lee, K.J., Park, C.W., Journal of Hazardous Materials, 369, 2019, 416-422. https://doi.org/10.1016/j.jhazmat.2019.02.054

[87] Zhao, L., Xie, S., Liu, Y., Liu, Q., Song, X., Li, X., Nanoscale, 11[38] 2019, 17831-17840. https://doi.org/10.1039/C9NR05503G

[88] Wang, S., Wu, N., Langmuir, 30[12] 2014, 3477-3486. https://doi.org/10.1021/la500182f

[89] Ge, Y., Wang, T., Zheng, M., Jiang, Z., Wang, S., Nanotechnology, 30[31] 2019, 315702. https://doi.org/10.1088/1361-6528/ab19c7

[90] Ren, M., Guo, W., Guo, H., Ren, X., ACS Applied Materials and Interfaces, 11[25] 2019, 22761-22767. https://doi.org/10.1021/acsami.9b05925

[91] Li, M., Li, D., Journal of Colloid and Interface Science, 532, 2018, 657-665. https://doi.org/10.1016/j.jcis.2018.08.034

[92] McNeill, J.M., Nama, N., Braxton, J.M., Mallouk, T.E., ACS Nano, 14[6] 2020, 7520-7528. https://doi.org/10.1021/acsnano.0c03311

[93] Xu, Z., Chen, M., Lee, H., Feng, S.P., Park, J.Y., Lee, S., Kim, J.T., ACS Applied Materials and Interfaces, 11[17] 2019, 15727-15732. https://doi.org/10.1021/acsami.9b00174

[94] Ji, Y., Lin, X., Wang, D., Zhou, C., Wu, Y., He, Q., Chemistry - an Asian Journal, 14[14] 2019, 2450-2455. https://doi.org/10.1002/asia.201801716

[95] Fisher, D.J., Materials Research Foundations, 77, 2020, 1-153.

[96] Wang, R., Guo, W., Li, X., Liu, Z., Liu, H., Ding, S., RSC Advances, 7[67] 2017, 42462-42467. https://doi.org/10.1039/C7RA08127H

[97] Tan, T.T.Y., Cham, J.T.M., Reithofer, M.R., Hor, T.S.A., Chin, J.M., Chemical Communications, 50[96] 2014, 15175-15178. https://doi.org/10.1039/C4CC06952H

[98] Gai, M., Frueh, J., Si, T., Hu, N., Sukhorukov, G.B., He, Q., Colloids and Surfaces A, 510, 2016, 113-121. https://doi.org/10.1016/j.colsurfa.2016.04.042

[99] Reddy, N.K., Clasen, C., Korea Australia Rheology Journal, 26[1] 2014, 73-79. https://doi.org/10.1007/s13367-014-0008-2

[100] Yariv, E., Crowdy, D., Physical Review Fluids, 5[11], 2020, 112001.
https://doi.org/10.1103/PhysRevFluids.5.112001

[101] Ma, X., Jang, S., Popescu, M.N., Uspal, W.E., Miguel-López, A., Hahn, K., Kim,
D.P., Sánchez, S., ACS Nano, 10[9] 2016, 8751-8759.
https://doi.org/10.1021/acsnano.6b04358

[102] Eloul, S., Poon, W.C.K., Farago, O., Frenkel, D., Physical Review Letters, 124[18]
2020, 188001. https://doi.org/10.1103/PhysRevLett.124.188001

[103] Yang, M., Wysocki, A., Ripoll, M., Soft Matter, 10[33] 2014, 6208-6218.
https://doi.org/10.1039/C4SM00621F

[104] Premlata, A.R., Wei, H.H., Journal of Fluid Mechanics, 882, 2020, A7.
https://doi.org/10.1017/jfm.2019.780

[105] Ma, X., Sanchez, S., Chemical Communications, 51[25] 2015, 5467-5470.
https://doi.org/10.1039/C4CC08285K

[106] Moo, J.G.S., Wang, H., Pumera, M., Chemistry - a European Journal, 22[1] 2016,
355-360. https://doi.org/10.1002/chem.201503473

[107] Wu, Y., Si, T., Lin, X., He, Q., Chemical Communications, 51[3] 2015, 511-514.
https://doi.org/10.1039/C4CC07182D

[108] Wang, H., Sofer, Z., Eng, A.Y.S., Pumera, M., Chemistry - a European Journal,
20[46] 2014, 14946-14950. https://doi.org/10.1002/chem.201404238

[109] Pinchasik, B.E., Möhwald, H., Skirtach, A.G., Small, 10[13] 2014, 2670-2677.
https://doi.org/10.1002/smll.201303571

[110] Zhou, L., Zhang, H., Bao, H., Wei, Y., Fu, H., Cai, W., ACS Applied Nano Materials,
3[1] 2020, 624-632. https://doi.org/10.1021/acsanm.9b02159

[111] Zhao, G., Pumera, M., Nanoscale, 6[19] 2014, 11177-11180.
https://doi.org/10.1039/C4NR02393E

[112] Gao, W., D'Agostino, M., Garcia-Gradilla, V., Orozco, J., Wang, J., Small, 9[3] 2013,
467-471. https://doi.org/10.1002/smll.201201864

[113] Gao, W., Pei, A., Wang, J., ACS Nano, 6[9] 2012, 8432-8438.
https://doi.org/10.1021/nn303309z

[114] Wu, Y., Wu, Z., Lin, X., He, Q., Li, J., ACS Nano, 6[12] 2012, 10910-10916.
https://doi.org/10.1021/nn304335x

[115] Fattah, Z., Loget, G., Lapeyre, V., Garrigue, P., Warakulwit, C., Limtrakul, J.,
Bouffier, L., Kuhn, A., Electrochimica Acta, 56[28] 2011, 10562-10566.
https://doi.org/10.1016/j.electacta.2011.01.048

[116] Ebbens, S.J., Howse, J.R., Langmuir, 27[20] 2011, 12293-12296.
https://doi.org/10.1021/la2033127

[117] Moo, J.G.S., Pumera, M., ACS Sensors, 1[7] 2016, 949-957.
https://doi.org/10.1021/acssensors.6b00314

[118] Wei, L., Ke, L., Yi-Xi, C., Xiao-Yu, H., Jun-Fu, W., Journal of Tianjin Polytechnic
University, 36[1] 2017, 41-47.

[119] Chen, C., Tang, S., Teymourian, H., Karshalev, E., Zhang, F., Li, J., Mou, F., Liang,
Y., Guan, J., Wang, J., Angewandte Chemie, 57[27] 2018, 8110-8114.
https://doi.org/10.1002/anie.201803457

[120] Bregulla, A.P., Yang, H., Cichos, F., ACS Nano, 8[7] 2014, 6542-6550.
https://doi.org/10.1021/nn501568e

[121] Qian, B., Montiel, D., Bregulla, A., Cichos, F., Yang, H., Chemical Science, 4[4]
2013, 1420-1429. https://doi.org/10.1039/c2sc21263c

[122] Buttinoni, I., Volpe, G., Kümmel, F., Volpe, G., Bechinger, C., Journal of Physics -
Condensed Matter, 24[28] 2012, 284129. https://doi.org/10.1088/0953-
8984/24/28/284129

[123] Samin, S., Van Roij, R., Physical Review Letters, 115[18] 2015, 188305.
https://doi.org/10.1103/PhysRevLett.115.188305

[124] Wu, Y., Si, T., Shao, J., Wu, Z., He, Q., Nano Research, 9[12] 2016, 3747-3756.
https://doi.org/10.1007/s12274-016-1245-0

[125] Ilic, O., Kaminer, I., Lahini, Y., Buljan, H., Soljačić, M., ACS Photonics, 3[2] 2016,
197-202. https://doi.org/10.1021/acsphotonics.5b00605

[126] Io, C.W., Chen, T.Y., Yeh, J.W., Cai, S.C., Physical Review E, 96[6] 2017, 062601.
https://doi.org/10.1103/PhysRevE.96.062601

[127] Pourrahimi, A.M., Villa, K., Sofer, Z., Pumera, M., Small Methods, 3[11] 2019,
1900258. https://doi.org/10.1002/smtd.201900258

[128] Pourrahimi, A.M., Villa, K., Palenzuela, C.L.M., Ying, Y., Sofer, Z., Pumera, M.,
Advanced Functional Materials, 29[22] 2019, 1808678.
https://doi.org/10.1002/adfm.201808678

[129] Dong, R., Hu, Y., Wu, Y., Gao, W., Ren, B., Wang, Q., Cai, Y., Journal of the
American Chemical Society, 139[5] 2017, 1722-1725.
https://doi.org/10.1021/jacs.6b09863

[130] Wang, X., Baraban, L., Misko, V.R., Nori, F., Huang, T., Cuniberti, G., Fassbender,
J., Makarov, D., Small, 14[44] 2018, 1802537. https://doi.org/10.1002/smll.201802537

[131] Maric, T., Nasir, M.Z.M., Webster, R.D., Pumera, M., Advanced Functional Materials, 30[9] 2020, 1908614. https://doi.org/10.1002/adfm.201908614

[132] Vutukuri, H.R., Bet, B., Van Roij, R., Dijkstra, M., Huck, W.T.S., Scientific Reports, 7[1] 2017, 16758. https://doi.org/10.1038/s41598-017-16731-5

[133] Sridhar, V., Park, B.W., Sitti, M., Advanced Functional Materials, 28[25] 2018, 1704902. https://doi.org/10.1002/adfm.201704902

[134] Wang, Q.L., Wang, C., Dong, R.F., Pang, Q.Q., Cai, Y.P., Inorganic Chemistry Communications, 91, 2018, 1-4. https://doi.org/10.1016/j.inoche.2018.02.020

[135] Dong, R., Zhang, Q., Gao, W., Pei, A., Ren, B., ACS Nano, 10[1] 2016, 839-844. https://doi.org/10.1021/acsnano.5b05940

[136] Gao, Y.R., Mou, F.Z., Xu, L.L., Guan, J.G., Journal of Wuhan University of Technology, 38[12] 2016, 8-13.

[137] Mou, F., Kong, L., Chen, C., Chen, Z., Xu, L., Guan, J., Nanoscale, 8[9] 2016, 4976-4983. https://doi.org/10.1039/C5NR06774J

[138] Li, Y., Mou, F., Chen, C., You, M., Yin, Y., Xu, L., Guan, J., RSC Advances, 6[13] 2016, 10697-10703. https://doi.org/10.1039/C5RA26798F

[139] Singh, D.P., Choudhury, U., Fischer, P., Mark, A.G., Advanced Materials, 29[32] 2017, 1701328. https://doi.org/10.1002/adma.201701328

[140] Wu, Y., Dong, R., Zhang, Q., Ren, B., Nano-Micro Letters, 9[3] 2017, 30. https://doi.org/10.1007/s40820-017-0133-9

[141] Zhang, Q., Dong, R., Wu, Y., Gao, W., He, Z., Ren, B., ACS Applied Materials and Interfaces, 9[5] 2017, 4674-4683. https://doi.org/10.1021/acsami.6b12081

[142] Uspal, W.E., Journal of Chemical Physics, 150[11] 2019, 114903. https://doi.org/10.1063/1.5080967

[143] Dai, B., Wang, J., Xiong, Z., Zhan, X., Dai, W., Li, C.C., Feng, S.P., Tang, J., Nature Nanotechnology, 11[12] 2016, 1087-1092. https://doi.org/10.1038/nnano.2016.187

[144] Wang, Q., Dong, R., Wang, C., Xu, S., Chen, D., Liang, Y., Ren, B., Gao, W., Cai, Y., ACS Applied Materials and Interfaces, 11[6] 2019, 6201-6207. https://doi.org/10.1021/acsami.8b17563

[145] Patino, T., Arqué, X., Mestre, R., Palacios, L., Sánchez, S., Accounts of Chemical Research, 51[11] 2018, 2662-2671. https://doi.org/10.1021/acs.accounts.8b00288

[146] Falasco, G., Pfaller, R., Bregulla, A.P., Cichos, F., Kroy, K., Physical Review E, 94[3] 2016, 030602. https://doi.org/10.1103/PhysRevE.94.030602

[147] Patiño, T., Feiner-Gracia, N., Arqué, X., Miguel-López, A., Jannasch, A., Stumpp, T., Schäffer, E., Albertazzi, L., Sánchez, S., Journal of the American Chemical Society, 140[25] 2018, 7896-7903. https://doi.org/10.1021/jacs.8b03460

[148] Schattling, P., Thingholm, B., Städler, B., Chemistry of Materials, 27[21] 2015, 7412-7418. https://doi.org/10.1021/acs.chemmater.5b03303

[149] Schattling, P.S., Ramos-Docampo, M.A., Salgueiriño, V., Städler, B., ACS Nano, 11[4] 2017, 3973-3983. https://doi.org/10.1021/acsnano.7b00441

[150] Ma, X., Sánchez, S., Tetrahedron, 73[33] 2017, 4883-4886. https://doi.org/10.1016/j.tet.2017.06.048

[151] Ma, X., Wang, X., Hahn, K., Sánchez, S., ACS Nano, 10[3] 2016, 3597-3605. https://doi.org/10.1021/acsnano.5b08067

[152] Ma, X., Jannasch, A., Albrecht, U.R., Hahn, K., Miguel-López, A., Schäffer, E., Sánchez, S., Nano Letters, 15[10] 2015. 7043-7050. https://doi.org/10.1021/acs.nanolett.5b03100

[153] Hu, Y., Sun, Y., Biochemical Engineering Journal, 149, 2019, 107242. https://doi.org/10.1016/j.bej.2019.107242

[154] Liu, J., Guo, H.L., Li, Z.Y., Nanoscale, 8[47] 2016, 19894-19900. https://doi.org/10.1039/C6NR07470G

[155] Nedev, S., Carretero-Palacios, S., Kühler, P., Lohmüller, T., Urban, A.S., Anderson, L.J.E., Feldmann, J., ACS Photonics, 2[4] 2015, 491-496. https://doi.org/10.1021/ph500371z

[156] Olarte-Plata, J.D., Bresme, F., The Journal of Chemical Physics, 152[20] 2020, 204902. https://doi.org/10.1063/5.0008237

[157] Feldmann, D., Arya, P., Lomadze, N., Kopyshev, A., Santer, S., Applied Physics Letters, 115[26] 2019, 263701. https://doi.org/10.1063/1.5129238

[158] Choudhary, A., Renganathan, T., Pushpavanam, S., Journal of Fluid Mechanics, 2020, A4, in press.

[159] Ginot, F., Solon, A., Kafri, Y., Ybert, C., Tailleur, J., Cottin-Bizonne, C., New Journal of Physics, 20[11] 2018, 115001. https://doi.org/10.1088/1367-2630/aae732

[160] Kessler, R., Bräuer, D., Dreissigacker, C., Drescher, J., Lozano, C., Bechinger, C., Born, P., Voigtmann, T., Review of Scientific Instruments, 91[1] 2020, 013902. https://doi.org/10.1063/1.5124895

[161] Sridhar, V., Podjaski, F., Kröger, J., Jiménez-Solano, A., Park, B.W., Lotsch, B.V., Sitti, M., Proceedings of the National Academy of Sciences of the United States of America, 117[40] 2020, 24748-24756. https://doi.org/10.1073/pnas.2007362117

[162] Hasnain, J., Menzl, G., Jungblut, S., Dellago, C., Soft Matter, 13[5] 2017, 930-936. https://doi.org/10.1039/C6SM01898J

[163] Khalil, I.S.M., Magdanz, V., Sanchez, S., Schmidt, O.G., Misra, S., International Journal of Advanced Robotic Systems, 12[1] 2015, 1-7. https://doi.org/10.5772/58873

[164] Li, T., Zhang, A., Shao, G., Wei, M., Guo, B., Zhang, G., Li, L., Wang, W., Advanced Functional Materials, 28[25] 2018, 1706066. https://doi.org/10.1002/adfm.201706066

[165] Aizawa, S., Seto, K., Tokunaga, E., Applied Sciences, 8[4] 2018, 653. https://doi.org/10.3390/app8040653

[166] Fei, W., Driscoll, M.M., Chaikin, P.M., Bishop, K.J.M., Soft Matter, 14[23] 2018, 4661-4665. https://doi.org/10.1039/C8SM00518D

[167] Yu, S., Ma, N., Yu, H., Sun, H., Chang, X., Wu, Z., Deng, J., Zhao, S., Wang, W., Zhang, G., Zhang, W., Zhao, Q., Li, T., Nanomaterials, 9[12] 2019, 1672. https://doi.org/10.3390/nano9121672

[168] Mair, L.O., Evans, B., Hall, A.R., Carpenter, J., Shields, A., Ford, K., Millard, M., Superfine, R., Journal of Physics D, 44[12] 2011, 125001. https://doi.org/10.1088/0022-3727/44/12/125001

[169] Chen, Y.L., Jiang, H.R., Applied Physics Letters, 109[19] 2016, 191605. https://doi.org/10.1063/1.4967740

[170] Nishiguchi, D., Iwasawa, J., Jiang, H.R., Sano, M., New Journal of Physics, 20[1] 2018, 015002. https://doi.org/10.1088/1367-2630/aa9b48

[171] Sindoro, M., Granick, S., Angewandte Chemie, 57[51] 2018, 16773-16776. https://doi.org/10.1002/anie.201810862

[172] Shklyaev, S., EPL, 110[5] 2015, 54002. https://doi.org/10.1209/0295-5075/110/54002

[173] Ekanem, E.E., Zhang, Z., Vladisavljević, G.T., Langmuir, 33[34] 2017, 8476-8482. https://doi.org/10.1021/acs.langmuir.7b02506

[174] Nisisako, T., Current Opinion in Colloid and Interface Science, 25, 2016, 1-12. https://doi.org/10.1016/j.cocis.2016.05.003

[175] Hessberger, T., Braun, L.B., Henrich, F., Müller, C., Giesselmann, F., Serra, C., Zentel, R., Journal of Materials Chemistry C, 4[37] 2016, 8778-8786. https://doi.org/10.1039/C6TC03378D

[176] Sun, X.T., Yang, C.G., Xu, Z.R., RSC Advances, 6[15] 2016, 12042-12047. https://doi.org/10.1039/C5RA24531A

[177] Bormashenko, E., Bormashenko, Y., Pogreb, R., Gendelman, O., Langmuir, 27[1] 2011, 7-10. https://doi.org/10.1021/la103653p

[178] Zhang, L., Xiao, Z., Chen, X., Chen, J., Wang, W., ACS Nano, 13[8] 2019, 8842-8853. https://doi.org/10.1021/acsnano.9b02100

[179] Wu, Y., Fu, A., Yossifon, G., Small, 16[22] 2020, 1906682. https://doi.org/10.1002/smll.201906682

[180] Bastos-Arrieta, J., Bauer, C., Eychmüller, A., Simmchen, J., Journal of Chemical Physics, 150[14] 2019, 144902. https://doi.org/10.1063/1.5085838

[181] Yu, T., Chuphal, P., Thakur, S., Reigh, S.Y., Singh, D.P., Fischer, P., Chemical Communications, 54[84] 2018, 11933-11936. https://doi.org/10.1039/C8CC06467A

[182] Fernández-Medina, M., Qian, X., Hovorka, O., Städler, B., Nanoscale, 11[2] 2019, 733-741. https://doi.org/10.1039/C8NR08071B

[183] Boymelgreen, A., Yossifon, G., Miloh, T., Langmuir, 32[37] 2016, 9540-9547. https://doi.org/10.1021/acs.langmuir.6b01758

[184] Shklyaev, S., Brady, J.F., Córdova-Figueroa, U.M., Journal of Fluid Mechanics, 748, 2014, 488-520. https://doi.org/10.1017/jfm.2014.177

[185] Jo, I., Huang, Y., Zimmermann, W., Kanso, E., Physical Review E, 94[6] 2016, 063116. https://doi.org/10.1103/PhysRevE.94.063116

[186] Laumann, M., Förtsch, A., Kanso, E., Zimmermann, W., New Journal of Physics, 21[7] 2019, 073012. https://doi.org/10.1088/1367-2630/ab240c

[187] Kurzthaler, C., Devailly, C., Arlt, J., Franosch, T., Poon, W.C.K., Martinez, V.A., Brown, A.T., Physical Review Letters, 121[7] 2018, 078001. https://doi.org/10.1103/PhysRevLett.121.078001

[188] Lozano, C., Gomez-Solano, J.R., Bechinger, C., New Journal of Physics, 20[1] 2018, 015008. https://doi.org/10.1088/1367-2630/aa9ed1

[189] Mano, T., Delfau, J.B., Iwasawa, J., Sano, M., Proceedings of the National Academy of Sciences, 114[13] 2017, E2580-E2589. https://doi.org/10.1073/pnas.1616013114

[190] Ebbens, S., Jones, R.A.L., Ryan, A.J., Golestanian, R., Howse, J.R., Physical Review E, 82[1] 2010, 015304. https://doi.org/10.1103/PhysRevE.82.015304

[191] Rogowski, L.W., Zhang, X., Huang, L., Bhattacharjee, A., Lee, J.S., Becker, A.T., Kim, M.J., Proceedings of the IEEE International Conference on Robotics and Automation, 2019, 1352-1357.

[192] Dong, R., Li, J., Rozen, I., Ezhilan, B., Xu, T., Christianson, C., Gao, W., Saintillan, D., Ren, B., Wang, J., Scientific Reports, 5, 2015, 13226. https://doi.org/10.1038/srep13226

[193] Peng, X., Chen, Z., Kollipara, P.S., Liu, Y., Fang, J., Lin, L., Zheng, Y., Light: Science and Applications, 9[1] 2020, 141. https://doi.org/10.1038/s41377-020-00378-5

[194] Montenegro-Johnson, T.D., Physical Review Fluids, 3[6] 2018, 062201.
https://doi.org/10.1103/PhysRevFluids.3.062201

[195] Semenov, S.N., Schimpf, M.E., Journal of Physical Chemistry B, 124[29] 2020, 6398-6403. https://doi.org/10.1021/acs.jpcb.0c02258

[196] Kierulf, A., Azizi, M., Eskandarloo, H., Whaley, J., Liu, W., Perez-Herrera, M., You, Z., Abbaspourrad, A., Food Hydrocolloids, 91, 2019, 301-310.
https://doi.org/10.1016/j.foodhyd.2019.01.037

[197] Zheng, X., Ten Hagen, B., Kaiser, A., Wu, M., Cui, H., Silber-Li, Z., Löwen, H., Physical Review E, 88[3] 2013, 032304. https://doi.org/10.1103/PhysRevE.88.032304

[198] Ghosh, P.K., Misko, V.R., Marchesoni, F., Nori, F., Physical Review Letters, 110[26] 2013, 268301. https://doi.org/10.1103/PhysRevLett.110.268301

[199] Campbell, A.I., Ebbens, S.J., Illien, P., Golestanian, R., Nature Communications, 10[1] 2019, 3952. https://doi.org/10.1038/s41467-019-11842-1

[200] Liu, Z., Du, L., Guo, W., Mei, D.C., European Physical Journal B, 89[10] 2016, 222.
https://doi.org/10.1140/epjb/e2016-70453-3

[201] Wei, H.H., Jan, J.S., Journal of Fluid Mechanics, 657, 2010, 64-88.

[202] Araki, T., Maciołek, A., Soft Matter, 15[26] 2019, 5243-5254.
https://doi.org/10.1017/S0022112010001369

[203] Araki, T., Fukai, S., Soft Matter, 11[17] 2015, 3470-3479.
https://doi.org/10.1039/C9SM00509A

[204] Bickel, T., Majee, A., Würger, A., Physical Review E, 88[1] 2013, 012301.
https://doi.org/10.1103/PhysRevE.88.012301

[205] Huang, M.J., Schofield, J., Kapral, R., Soft Matter, 12[25] 2016, 5581-5589.
https://doi.org/10.1039/C6SM00830E

[206] Speck, T., EPL, 123[2] 2018, 20007. https://doi.org/10.1209/0295-5075/123/20007

[207] Majee, A., European Physical Journal E, 40[3] 2017, 30.
https://doi.org/10.1140/epje/i2017-11518-4

[208] Chen, L., Mo, C., Wang, L., Cui, H., Microfluidics and Nanofluidics, 23[5] 2019, 73.
https://doi.org/10.1007/s10404-019-2230-1

[209] Burelbach, J., Stark, H., Physical Review E, 100[4] 2019, 042612.
https://doi.org/10.1103/PhysRevE.100.042612

[210] Natale, G., Datt, C., Hatzikiriakos, S.G., Elfring, G.J., Physics of Fluids, 29[12] 2017, 123102. https://doi.org/10.1063/1.5002729

[211] Gomez-Solano, J.R., Blokhuis, A., Bechinger, C., Physical Review Letters, 116[13] 2016, 138301. https://doi.org/10.1103/PhysRevLett.116.138301

[212] Chakraborty, D., Journal of Chemical Physics, 149[17] 2018, 174907. https://doi.org/10.1063/1.5046059

[213] Wang, X., In, M., Blanc, C., Malgaretti, P., Nobili, M., Stocco, A., Faraday Discussions, 191, 2016, 305-324. https://doi.org/10.1039/C6FD00025H

[214] Debnath, T., Ghosh, P.K., Li, Y., Marchesoni, F., Nori, F., Journal of Chemical Physics, 150[15] 2019, 154902. https://doi.org/10.1063/1.5081125

[215] Hauke, F., Löwen, H., Liebchen, B., Journal of Chemical Physics, 152[1] 2020, 014903. https://doi.org/10.1063/1.5128641

[216] Chen, C., Chang, X., Teymourian, H., Ramírez-Herrera, D.E., Esteban-Fernández de Ávila, B., Lu, X., Li, J., He, S., Fang, C., Liang, Y., Mou, F., Guan, J., Wang, J., Angewandte Chemie, 57[1] 2018, 241-245. https://doi.org/10.1002/anie.201710376

[217] Biswas, B., Manna, R.K., Laskar, A., Kumar, P.B.S., Adhikari, R., Kumaraswamy, G., ACS Nano, 11[10] 2017, 10025-10031. https://doi.org/10.1021/acsnano.7b04265

[218] Liebchen, B., Löwen, H., Accounts of Chemical Research, 51[12] 2018, 2982-2990. https://doi.org/10.1021/acs.accounts.8b00215

[219] Ji, Y., Lin, X., Wu, Z., Wu, Y., Gao, W., He, Q., Angewandte Chemie, 58[35] 2019, 12200-12205. https://doi.org/10.1002/anie.201907733

[220] Popescu, M.N., Uspal, W.E., Bechinger, C., Fischer, P., Nano Letters, 18[9] 2018, 5345-5349. https://doi.org/10.1021/acs.nanolett.8b02572

[221] Nejad, M.R., Najafi, A., Soft Matter, 15[15] 2019, 3248-3255. https://doi.org/10.1039/C9SM00058E

[222] Huang, Z., Chen, P., Zhu, G., Yang, Y., Xu, Z., Yan, L.T., ACS Nano, 12[7] 2018, 6725-6735. https://doi.org/10.1021/acsnano.8b01842

[223] Wang, J., Huang, M.J., Kapral, R., Journal of Chemical Physics, 153[1] 2020, 014902. https://doi.org/10.1063/5.0012265

[224] Schmieding, L.C., Lauga, E., Montenegro-Johnson, T.D., Physical Review Fluids, 2[3] 2017, 034201. https://doi.org/10.1103/PhysRevFluids.2.034201

[225] Campbell, A.I., Ebbens, S.J., Langmuir, 29[46] 2013, 14066-14073. https://doi.org/10.1021/la403450j

[226] Campbell, A.I., Wittkowski, R., Ten Hagen, B., Löwen, H., Ebbens, S.J., Journal of Chemical Physics, 147[8] 2017, 084905. https://doi.org/10.1063/1.4998605

[227] Das, S., Garg, A., Campbell, A.I., Howse, J., Sen, A., Velegol, D., Golestanian, R., Ebbens, S.J., Nature Communications, 6, 2015, 8999. https://doi.org/10.1038/ncomms9999

[228] Uspal, W.E., Popescu, M.N., Dietrich, S., Tasinkevych, M., Soft Matter, 11[3] 2015, 434-438. https://doi.org/10.1039/C4SM02317J

[229] Rosenthal, G., Klapp, S.H.L., Journal of Chemical Physics, 134[15] 2011, 154707. https://doi.org/10.1063/1.3579453

[230] Rashidi, A., Razavi, S., Wirth, C.L., Physical Review E, 101[4] 2020, 042606. https://doi.org/10.1103/PhysRevE.101.042606

[231] Boymelgreen, A., Yossifon, G., Langmuir, 31[30] 2015, 8243-8250. https://doi.org/10.1021/acs.langmuir.5b01199

[232] Spagnolie, S.E., Lauga, E., Journal of Fluid Mechanics, 700, 2012, 105-147. https://doi.org/10.1017/jfm.2012.101

[233] Borówko, M., Pöschel, T., Sokołowski, S., Staszewski, T., Journal of Physical Chemistry B, 117[4] 2013, 1166-1175. https://doi.org/10.1021/jp3105979

[234] Ghosh, P.K., Hänggi, P., Marchesoni, F., Nori, F., Physical Review E, 89[6] 2014, 062115. https://doi.org/10.1103/PhysRevE.89.062115

[235] Ghosh, P.K., Journal of Chemical Physics, 141[6] 2014, 061102. https://doi.org/10.1063/1.4892970

[236] Ao, X., Ghosh, P.K., Li, Y., Schmid, G., Hänggi, P., Marchesoni, F., European Physical Journal - Special Topics, 223[14] 2014, 3227-3242. https://doi.org/10.1140/epjst/e2014-02329-1

[237] Ao, X., Ghosh, P.K., Li, Y., Schmid, G., Hänggi, P., Marchesoni, F., European Physical Journal - Special Topics, 223[14] 2014, 3227-3242. https://doi.org/10.1140/epjst/e2014-02329-1

[238] Li, Y., Ghosh, P.K., Marchesoni, F., Li, B., Physical Review E, 90[6] 2014, 062301. https://doi.org/10.1103/PhysRevE.90.062301

[239] Ao, X., Ghosh, P.K., Li, Y., Schmid, G., Hänggi, P., Marchesoni, F., EPL, 109[1] 2015, 10003. https://doi.org/10.1209/0295-5075/109/10003

[240] Levis, D., Liebchen, B., Physical Review E, 100[1] 2019, 012406. https://doi.org/10.1103/PhysRevE.100.012406

[241] Wang, X., In, M., Blanc, C., Nobili, M., Stocco, A., Soft Matter, 11[37] 2015, 7376-7384. https://doi.org/10.1039/C5SM01111F

[242] Mangal, R., Nayani, K., Kim, Y.K., Bukusoglu, E., Córdova-Figueroa, U.M., Abbott, N.L., Langmuir, 33[41] 2017, 10917-10926. https://doi.org/10.1021/acs.langmuir.7b02246

[243] Ibrahim, Y., Liverpool, T.B., EPL, 111[4] 2015, 48008. https://doi.org/10.1209/0295-5075/111/48008

[244] Cui, H.H., Tan, X.J., Zhang, H.Y., Chen, L., Acta Physica Sinica, 64[13] 2015, 134705.

[245] Yu, H., Kopach, A., Misko, V.R., Vasylenko, A.A., Makarov, D., Marchesoni, F., Nori, F., Baraban, L., Cuniberti, G., Small, 12[42] 2016, 5882-5890. https://doi.org/10.1002/smll.201602039

[246] Harder, J., Cacciuto, A., Physical Review E, 97[2] 2018, 022603. https://doi.org/10.1103/PhysRevE.97.022603

[247] Bickel, T., Zecua, G., Würger, A., Physical Review E, 89[5] 2014, 050303. https://doi.org/10.1103/PhysRevE.89.050303

[248] Traverso, T., Michelin, S., Physical Review Fluids, 5[10] 2020, 104203. https://doi.org/10.1103/PhysRevFluids.5.104203

[249] Guzmán-Lastra, F., Kaiser, A., Löwen, H., Nature Communications, 7, 2016, 13519. https://doi.org/10.1038/ncomms13519

[250] Choudhury, U., Straube, A.V., Fischer, P., Gibbs, J.G., Höfling, F., New Journal of Physics, 19[12] 2017, 125010. https://doi.org/10.1088/1367-2630/aa9b4b

[251] Popescu, M.N., Uspal, W.E., Dietrich, S., Journal of Physics - Condensed Matter, 29[13] 2017, 134001. https://doi.org/10.1088/1361-648X/aa5bf1

[252] Katuri, J., Caballero, D., Voituriez, R., Samitier, J., Sanchez, S., ACS Nano, 12[7] 2018, 7282-7291. https://doi.org/10.1021/acsnano.8b03494

[253] Uspal, W.E., Popescu, M.N., Tasinkevych, M., Dietrich, S., New Journal of Physics, 20[1] 2018, 015013. https://doi.org/10.1088/1367-2630/aa9f9f

[254] Katuri, J., Uspal, W.E., Simmchen, J., Miguel-López, A., Sánchez, S., Science Advances, 4[1] 2018, eaao1755. https://doi.org/10.1126/sciadv.aao1755

[255] Si, B.R., Patel, P., Mangal, R., Langmuir, 36[40] 2020, 11888-11898. https://doi.org/10.1021/acs.langmuir.0c01924

[256] Uspal, W.E., Popescu, M.N., Dietrich, S., Tasinkevych, M., Journal of Chemical Physics, 150[20] 2019, 204904. https://doi.org/10.1063/1.5091760

[257] Issa, M.W., Baumgartner, N.R., Kalil, M.A., Ryan, S.D., Wirth, C.L., ACS Omega, 4[8] 2019, 13034-13041. https://doi.org/10.1021/acsomega.9b00765

[258] Razavi, S., Hernandez, L.M., Read, A., Vargas, W.L., Kretzschmar, I., Journal of Colloid and Interface Science, 558, 2019, 95-99. https://doi.org/10.1016/j.jcis.2019.09.084

[259] Knapp, E.M., Dagastine, R.R., Tu, R.S., Kretzschmar, I., ACS Applied Materials and Interfaces, 12[4] 2020, 5128-5135. https://doi.org/10.1021/acsami.9b21067

[260] Debnath, T., Li, Y., Ghosh, P.K., Marchesoni, F., Physical Review E, 97[4] 2018, 042602. https://doi.org/10.1103/PhysRevE.97.042602

[261] Bayati, P., Najafi, A., Journal of Chemical Physics, 144[13] 2016, 134901. https://doi.org/10.1063/1.4944988

[262] Ishimoto, K., Gaffney, E.A., Physical Review E, 88[6] 2013, 062702. https://doi.org/10.1103/PhysRevE.88.062702

[263] Yang, L., Wang, T., Yang, X., Jiang, G., Luckham, P.F., Xu, J., Li, X., Ni, X., Industrial and Engineering Chemistry Research, 58[23] 2019, 9795-9805. https://doi.org/10.1021/acs.iecr.9b01714

[264] Song, G., Chen, M., Zhang, Y., Cui, L., Qu, H., Zheng, X., Wintermark, M., Liu, Z., Rao, J., Nano Letters, 18[1] 2018, 182-189. https://doi.org/10.1021/acs.nanolett.7b03829

[265] Sanchez-Vazquez, B., Amaral, A.J.R., Yu, D.G., Pasparakis, G., Williams, G.R., AAPS PharmSciTech, 18[5] 2017, 1460-1468. https://doi.org/10.1208/s12249-016-0638-4

[266] Kilinc, D., Lesniak, A., Rashdan, S.A., Gandhi, D., Blasiak, A., Fannin, P.C., von Kriegsheim, A., Kolch, W., Lee, G.U., Advanced Healthcare Materials, 4[3] 2015, 395-404. https://doi.org/10.1002/adhm.201400391

[267] Zhou, W., Sun, W., Yang, P., Progress in Chemistry, 30[11] 2018, 1601-1614.

[268] Pacheco, M., Jurado-Sánchez, B., Escarpa, A., Chemical Science, 9[42] 2018, 8056-8064. https://doi.org/10.1039/C8SC03681K

[269] Jurado-Sánchez, B., Pacheco, M., Rojo, J., Escarpa, A., Angewandte Chemie, 56[24] 2017, 6957-6961. https://doi.org/10.1002/anie.201701396

[270] Pacheco, M., Jurado-Sánchez, B., Escarpa, A., Analytical Chemistry, 90[4] 2018, 2912-2917. https://doi.org/10.1021/acs.analchem.7b05209

[271] Kaewsaneha, C., Tangboriboonrat, P., Polpanich, D., Eissa, M., Elaissari, A., Colloids and Surfaces A, 439, 2013, 35-42. https://doi.org/10.1016/j.colsurfa.2013.01.004

[272] Yang, Q., Miao, X., Loos, K., Macromolecular Chemistry and Physics, 219[19] 2018, 1800267. https://doi.org/10.1002/macp.201800267

[273] Wang, C.Y., Yang, R., Chen, Y.H., Tong, Z., Chemical Journal of Chinese Universities, 31[5] 2010, 864-866.

[274] Isenbügel, K., Gehrke, Y., Ritter, H., Macromolecular Rapid Communications, 33[1] 2012, 41-46. https://doi.org/10.1002/marc.201100499

[275] Zenerino, A., Peyratout, C., Aimable, A., Journal of Colloid and Interface Science, 450, 2015, 174-181. https://doi.org/10.1016/j.jcis.2015.03.011

[276] Kirillova, A., Schliebe, C., Stoychev, G., Jakob, A., Lang, H., Synytska, A., ACS Applied Materials and Interfaces, 7[38] 2015, 21224-21225. https://doi.org/10.1021/acsami.5b05224

[277] Panwar, K., Jassal, M., Agrawal, A.K., RSC Advances, 6[95] 2016, 92754-92764. https://doi.org/10.1039/C6RA12378C

[278] Shi, Y., Zhang, Q., Liu, Y., Chang, J., Guo, J., Chinese Journal of Catalysis, 40[5] 2019, 786-794. https://doi.org/10.1016/S1872-2067(19)63332-2

[279] Ruhland, T.M., McKenzie, H.S., Skelhon, T.S., Bon, S.A.F., Walther, A., Müller, A.H.E., Polymer, 79, 2015, 299-308. https://doi.org/10.1016/j.polymer.2015.10.022

[280] Zhao, R., Han, T., Sun, D., Huang, L., Liang, F., Liu, Z., Langmuir, 35[35] 2019, 11435-11442. https://doi.org/10.1021/acs.langmuir.9b01400

[281] Zhai, W., Wang, B., Wang, Y., He, Y.F., Song, P., Wang, R.M., Colloids and Surfaces A, 503, 2016, 94-100. https://doi.org/10.1016/j.colsurfa.2016.05.025

[282] Park, J.H., Han, N., Song, J.E., Cho, E.C., Macromolecular Rapid Communications, 38[3] 2017, 1600621. https://doi.org/10.1002/marc.201600621

[283] Kadam, R., Zilli, M., Maas, M., Rezwan, K., Particle and Particle Systems Characterization, 35[3] 2018, 1700332. https://doi.org/10.1002/ppsc.201700332

[284] Tan, J.S.J., Wong, S.L.Y., Chen, Z., Advanced Materials Interfaces, 7[4] 2020, 1901961. https://doi.org/10.1002/admi.201901961

[285] Gao, W., Feng, X., Pei, A., Gu, Y., Li, J., Wang, J., Nanoscale, 5[11] 2013, 4696-4700. https://doi.org/10.1039/c3nr01458d

[286] Li, J., Ji, F., Ng, D.H.L., Liu, J., Bing, X., Wang, P., Chemical Engineering Journal, 369, 2019, 611-620. https://doi.org/10.1016/j.cej.2019.03.101

[287] Wang, G., Wang, K., Wang, Y., Particuology, 41, 2018, 112-117. https://doi.org/10.1016/j.partic.2017.12.006

[288] Li, X., Mou, F., Guo, J., Deng, Z., Chen, C., Xu, L., Luo, M., Guan, J., Micromachines, 9[1] 2018, 23. https://doi.org/10.3390/mi9010023

[289] Burton, L.J., Cheng, N., Bush, J.W.M., Integrative and Comparative Biology, 54[6] 2014, 969-973. https://doi.org/10.1093/icb/icu052

[290] Lee, C.S., Gong, J., Oh, D.S., Jeon, J.R., Chang, Y.S., ACS Applied Nano Materials, 1[2] 2018, 768-776. https://doi.org/10.1021/acsanm.7b00223

[291] Lu, A.X., Liu, Y., Oh, H., Gargava, A., Kendall, E., Nie, Z., Devoe, D.L., Raghavan, S.R., ACS Applied Materials and Interfaces, 8[24] 2016, 15676-15683. https://doi.org/10.1021/acsami.6b01245

[292] Delezuk, J.A.M., Ramírez-Herrera, D.E., Esteban-Fernández de Ávila, B., Wang, J., Nanoscale, 9[6] 2017, 2195-2200. https://doi.org/10.1039/C6NR09799E

[293] Jia, R., Jiang, H., Jin, M., Wang, X., Huang, J., (2015) Food Research International, 78, 433-441. https://doi.org/10.1016/j.foodres.2015.08.035

[294] Vilela, D., Stanton, M.M., Parmar, J., Sánchez, S., ACS Applied Materials and Interfaces, 9[27] 2017, 22093-22100. https://doi.org/10.1021/acsami.7b03006

[295] Brown, A.T., Vladescu, I.D., Dawson, A., Vissers, T., Schwarz-Linek, J., Lintuvuori, J.S., Poon, W.C.K., Soft Matter, 12[1] 2015, 131-140. https://doi.org/10.1039/C5SM01831E

[296] Šimkus, R., Meškienė, R., Aučynaitė, A., Ledas, Ž., Baronas, R., Meškys, R., Royal Society Open Science, 5[5] 2018, 180008. https://doi.org/10.1098/rsos.180008

[297] Jurado-Sánchez, B., Sattayasamitsathit, S., Gao, W., Santos, L., Fedorak, Y., Singh, V.V., Orozco, J., Galarnyk, M., Wang, J., Small, 11[4] 2015, 499-506. https://doi.org/10.1002/smll.201402215

[298] Ebbens, S.J., Gregory, D.A., Accounts of Chemical Research, 51[9] 2018, 1931-1939. https://doi.org/10.1021/acs.accounts.8b00243

[299] Archer, R.J., Campbell, A.I., Ebbens, S.J., Soft Matter, 11[34] 2015, 6872-6880. https://doi.org/10.1039/C5SM01323B

[300] Huo, X., Wu, Y., Boymelgreen, A., Yossifon, G., Langmuir, 36[25] 2020, 6963-6970. https://doi.org/10.1021/acs.langmuir.9b03036

[301] Teo, W.Z., Zboril, R., Medrik, I., Pumera, M., Chemistry - a European Journal, 22[14] 2016, 4789-4793. https://doi.org/10.1002/chem.201504912

[302] Mou, F., Chen, C., Zhong, Q., Yin, Y., Ma, H., Guan, J., ACS Applied Materials and Interfaces, 6[12] 2014, 9897-9903. https://doi.org/10.1021/am502729y

[303] Wu, Z., Li, J., De Ávila, B.E.F., Li, T., Gao, W., He, Q., Zhang, L., Wang, J., Advanced Functional Materials, 25[48] 2015, 7497-7501. https://doi.org/10.1002/adfm.201503441

[304] Li, J., Shklyaev, O.E., Li, T., Liu, W., Shum, H., Rozen, I., Balazs, A.C., Wang, J., Nano Letters, 15[10] 2015, 7077-7085. https://doi.org/10.1021/acs.nanolett.5b03140

[305] Chen, C., Karshalev, E., Li, J., Soto, F., Castillo, R., Campos, I., Mou, F., Guan, J., Wang, J., ACS Nano, 10[11] 2016, 10389-10396. https://doi.org/10.1021/acsnano.6b06256

[306] Rojas, D., Jurado-Sánchez, B., Escarpa, A., Analytical Chemistry, 88[7] 2016, 4153-4160. https://doi.org/10.1021/acs.analchem.6b00574

[307] Uygun, D.A., Jurado-Sánchez, B., Uygun, M., Wang, J., Environmental Science: Nano, 3[3] 2016, 559-566. https://doi.org/10.1039/C6EN00043F

[308] Walther, A., Müller, A.H.E., Soft Matter, 4[4] 2008, 663-668. https://doi.org/10.1039/b718131k

[309] Pacheco, M., Jurado-Sánchez, B., Escarpa, A., Angewandte Chemie, 58[50] 2019, 18017-18024. https://doi.org/10.1002/anie.201910053

[310] Debnath, D., Ghosh, P.K., Misko, V.R., Li, Y., Marchesoni, F., Nori, F., Nanoscale, 12[17] 2020, 9717-9726. https://doi.org/10.1039/D0NR01765E

[311] Lee, K., Yi, Y., Yu, Y., Angewandte Chemie, 55[26] 2016, 7384-7387. https://doi.org/10.1002/anie.201601211

[312] Chen, B., Jia, Y., Gao, Y., Sanchez, L., Anthony, S.M., Yu, Y., ACS Applied Materials and Interfaces, 6[21] 2014, 18435-18439. https://doi.org/10.1021/am505510m

[313] Liebchen, B., Marenduzzo, D., Pagonabarraga, I., Cates, M.E., Physical Review Letters, 115[25] 2015, 258301. https://doi.org/10.1103/PhysRevLett.115.258301

[314] Xing, Y., Zhou, M., Du, X., Li, X., Li, J., Xu, T., Zhang, X., Applied Materials Today, 17, 2019, 85-91. https://doi.org/10.1016/j.apmt.2019.07.017

[315] Feng, Z.Q., Yan, K., Li, J., Xu, X., Yuan, T., Wang, T., Zheng, J., Materials Science and Engineering C, 104, 2019, 110001. https://doi.org/10.1016/j.msec.2019.110001

[316] Jiao, X., Wang, Z., Xiu, J., Dai, W., Zhao, L., Xu, T., Du, X., Wen, Y., Zhang, X., Applied Materials Today, 18, 2020, 100504. https://doi.org/10.1016/j.apmt.2019.100504

[317] Zhang, Z., Yan, H., Li, S., Liu, Y., Ran, P., Chen, W., Li, X., Chemical Engineering Journal, 404, 2021, 127073. https://doi.org/10.1016/j.cej.2020.127073

[318] Wu, Y., Lin, X., Wu, Z., Möhwald, H., He, Q., ACS Applied Materials and Interfaces, 6[13] 2014, 10476-10481. https://doi.org/10.1021/am502458h

[319] Yang, P.P., Zhai, Y.G., Qi, G.B., Lin, Y.X., Luo, Q., Yang, Y., Xu, A.P., Yang, C., Li, Y.S., Wang, L., Wang, H., Small, 12[39] 2016, 5423-5430. https://doi.org/10.1002/smll.201601965

[320] Xuan, M., Shao, J., Gao, C., Wang, W., Dai, L., He, Q., Angewandte Chemie, 57[38] 2018, 12463-12467. https://doi.org/10.1002/anie.201806759

[321] Ding, H.M., Ma, Y.Q., Nanoscale, 4[4] 2012, 1116-1122. https://doi.org/10.1039/C1NR11425E

[322] Panwar, K., Jassal, M., Agrawal, A.K., Carbohydrate Polymers, 187, 2018, 43-50. https://doi.org/10.1016/j.carbpol.2018.01.076

[323] Li, B., Wang, M., Chen, K., Cheng, Z., Chen, G., Zhang, Z., Macromolecular Rapid Communications, 36[12] 2015, 1200-1204. https://doi.org/10.1002/marc.201500063

[324] Simoncelli, S., Johnson, S., Kriegel, F., Lipfert, J., Feldmann, J., ACS Photonics, 4[11] 2017, 2843-2851. https://doi.org/10.1021/acsphotonics.7b00839

[325] Li, M., Brinkmann, M., Pagonabarraga, I., Seemann, R., Fleury, J.B., Communications Physics, 1[1] 2018, 23. https://doi.org/10.1038/s42005-018-0025-4

[326] Li, Q., Hu, E., Yu, K., Xie, R., Lu, F., Lu, B., Bao, R., Zhao, T., Dai, F., Lan, G., Advanced Functional Materials, 30[42] 2020, 2004153. https://doi.org/10.1002/adfm.202004153

[327] Stanton, M.M., Simmchen, J., Ma, X., Miguel-López, A., Sánchez, S., Advanced Materials Interfaces, 3[2] 2016, 1500505. https://doi.org/10.1002/admi.201500505

[328] Park, S., Yossifon, G., ACS Sensors, 5[4] 2020, 936-942. https://doi.org/10.1021/acssensors.9b02041

[329] Sakamoto, N., Hirai, Y., Onodera, T., Dezawa, T., Shibata, Y., Kasai, H., Oikawa, H., Yabu, H., (2019) Particle and Particle Systems Characterization, 36 (1), art. no. 1800311. https://doi.org/10.1002/ppsc.201800311

[330] María-Hormigos, R., Escarpa, A., Goudeau, B., Ravaine, V., Perro, A., Kuhn, A., Advanced Materials Interfaces, 7[10] 2020, 1902094. https://doi.org/10.1002/admi.201902094

[331] Cui, J., Long, D., Shapturenka, P., Kretzschmar, I., Chen, X., Wang, T., Colloids and Surfaces A, 513, 2017, 452-462. https://doi.org/10.1016/j.colsurfa.2016.11.017

[332] Wu, Y., Fu, A., Yossifon, G., Science Advances, 6[5] 2020, eaay4412. https://doi.org/10.1126/sciadv.aay4412

Materials Research Forum LLC
https://doi.org/10.21741/9781644901199